ANIMALS IN THE AMERICAN ECONOMY

Animals
in the
American
Economy

JOHN A. SIMS
&
LESLIE E. JOHNSON

[THE IOWA STATE UNIVERSITY PRESS]

Library of Congress Cataloging in Publication Data

Sims, John A.
    Animals in the American economy.

    Includes bibliographies.
    1. Domestic animals—United States. 2. Stock and stock-breeding—United States. I. Johnson,
Leslie Eckroat, 1905–1967, joint author. II. Title.
SF51.S48   636'.00973   77–83318
ISBN 0–8138–0245–8

© 1972 The Iowa State University Press
Ames, Iowa 50010. All rights reserved

Composed and printed by
The Iowa State University Press

First edition, 1972

Contents

Preface

THROUGHOUT the centuries man has found a need for his lesser relatives in the animal kingdom. In myriad ways animals of land, air, and sea have contributed to the cultural, spiritual, and economic well-being of mankind. Only a minute sample of the earth's animals has been domesticated and selectively bred by man for specific purposes. The distribution of these around the globe has taken place largely within the past century and is an on-going process.

Knowledge of these animal genetic resources and their interactions is still scanty even among animal scientists. This book attempts to familiarize the reader with the most important domesticated species, types, and breeds that have a place in the economy of the Americas. It is hoped that the information will prove intriguing to the casual observer as well as to those who will specialize in some phase of animal agriculture.

Much of the basic information presented herein on breeds of farm livestock was compiled in a manuscript by the late Dr. Leslie E. Johnson, head of the Iowa State University Animal Science Department, just prior to his death. This has been integrated and expanded into the present form.

The dynamic nature of breed development and improvement necessitates a continuing exploration of current literature to supplement this basic outline for those concerned with keeping abreast of new developments in the field. The literature is voluminous and readily available.

JOHN A. SIMS

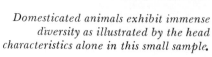

*Domesticated animals exhibit immense
diversity as illustrated by the head
characteristics alone in this small sample.*

*A*nd God blessed them, and God said unto them, Be fruitful, and multiply, and replenish the earth, and subdue it: and have dominion over the fish of the sea, and over the fowl of the air, and over every living thing that moveth upon the earth. *Genesis 1:28*

[CHAPTER ONE] Man's Dominion over Animals

THE GENETIC RESOURCES making up the animal kingdom are diverse almost beyond imagination. This living clay in man's hands has been revered throughout the centuries. The adjustments facing man and his animals in the late 20th century and beyond pose a more formidable task than in any previous era. How well has man measured up in his stewardship of animal resources?

The human race has been almost overwhelmingly successful in multiplying and replenishing the earth with its own kind. It has subdued large areas of the earth to the point of ravishment and has already driven several animal species to complete extinction. These drastic forces imposed on the earth's ecology give cause for concern, not only to the ecologist and conservationist but to the prudent man. Only in the fullness of time will it be known whether all the most useful genetic resources have been saved.

Certainly those species that have been nurtured and improved have vastly enhanced the welfare of those cultures which have fostered them. Furthermore, only a mere fraction of the potential contribution of domesticated animals has been realized. For example, the output of usable product per animal unit within a particular species may show a 100-fold difference between areas of nearly comparable environmental capability. Thus, increased knowledge of the species, types, and breeds of animals adapted to man's use is germane to all people as consumers of animal products and services and is essential to the practitioner of animal science.

The situation with respect to farm livestock was stated succinctly in a newspaper headline in the June 14, 1970, *Des Moines (Iowa) Register:* "Animal Scientists Race to Improve Livestock—the Consumer Wins." It is imperative to consumers in a complex and burgeoning human society that this race be won.

DOMESTICATION OF ANIMALS

In the early part of the period known as the New Stone Age, man won his greatest victory in his quest to control nature and develop his many talents—the discovery of the arts of crop production and animal husbandry. Once man discovered that a planted field and a tended herd yielded more food than nature at large with less of his labor than the hunting of wild plants and animals, his wanderings abated and his environment became more stable. He had more and better food with time to study and plan for things other than daily necessities. His health improved and his life-span was extended. This was the real beginning of modern civilization.

EARLY HISTORY. The discovery of the value of tended animals spurred man to domesticate many of the wild animals that roamed the world. By domestication we mean that the animals were tamed, bred, and developed so that man could care for them and use them in his everyday living. This was the beginning of the development of the farm animals that we use today for the production of food, fiber, work, and pleasure.

In general, present-day farm animals were domesticated long before written history began; thus the date of domestication is obscure. Bones and tools found buried at early campsites and drawings and carvings found in ancient caves indicate that it probably was in the early part of the New Stone Age (Neolithic period). This era extended over a period of 5,000–8,000 years and ended in what is known as the Bronze Age, which began 3,000 or more years before Christ. The area of earliest development was from Asia Minor eastward through northern Iraq toward the steppes of southern Asia.

Thus the farm and pleasure animals we work with today have undergone many generations of selection and development. They have long been bred to fulfill some need of man and not for survival of the fittest in occupying some niche within the animal world.

The methods used by early man to domesticate animals are even more obscure than the date of domestication. Hunters may have brought home young or crippled animals for pets or companions and thus learned how to care for them and use them. If many men in a camp did this, the camp would soon have possessed producing herds or flocks. There is evidence that early man used to drive wild animals into pits and closed canyons. One ancient Egyptian drawing shows hunters building a rock fence across the mouth of a narrow canyon to confine a group of wild animals. It is likely that most of these animals were slaughtered for food as needed, but some of the younger ones may have been held in the canyons and fed and tamed.

Feeding must have been used often to subdue and tame hungry

animals which man wanted to keep for his use. In following herds of wild animals for food, hunters probably often came upon weak and young animals they did not care to kill. Some of these may have been taken to good grassy areas near man's camps where predators feared to come. From such groups our early domesticated herds could have originated. Regardless of the methods used, it is certain that primitive man lived close to his animals and in many ways knew them better than we know our animals today.

The order in which man domesticated the different species and started using them is also indefinite. Our best authorities feel that the dog was probably the first animal to have been tamed by man. N. S. Shaler, an authority on early domestication, believes the dog was tamed for companionship and not for food, work, or sport. It was much later that the dog was bred and kept for these purposes. The ox, Asiatic buffalo, goat, and sheep were other early members of man's domesticated animals. It is known that at a very early date cattle, sheep, and goats were valued for their milk. The general use of animals for milk must have occurred only after many generations under domestication. Swine and birds attracted man's attention as food in an early era, and camels and elephants were used as beasts of burden during much of the Bronze Age.

There is considerable disagreement over the date of domestication of the horse. Some place the horse second to the dog; others feel that the early horse was hunted only for food and not tamed until well after the domestication of all the species named above. One authority cites evidence to show that not only was the horse domesticated late in the New Stone Age but he was not ridden by man until after he had been hitched to chariots.

A large part of the domestication of animals appears to have been done by the Aryan race in the Old World. Central and western Asia continually yield evidence that early man of this area used many species of tamed animals. It is not likely, however, that one area or one people produced all the domesticated animals. Swine appear to have been separately domesticated in the areas of the Alps, the Baltic Sea, and China. Chickens and elephants were most certainly first tamed in India. The alpaca, llama, guinea pig, and turkey were domesticated in the Americas since they did not occur in other areas. As man traveled in his migrations and wars, animals from different areas must have been crossed. Thus it seems likely that most of our farm animals of today trace to the blending of different types and possibly species which were separately domesticated. Such crosses would have had the genetic variation so desirable in a selection program where large changes in animals were desired.

Early man must have attempted to domesticate other species but failed. Reasons for such failures were probably lack of adaptability of the species to man's way of life and/or man's lack of a real need for such types of animal.

It is difficult to know when an animal passes from the wild state to the tame. Some species of animals are still in the process of being domesticated. Most authorities would classify the fox, mink, and chinchilla as being in the process of domestication. With our present rapid expansion of human population and reduction of wildlife, it appears we should give thought to the domestication of many other species of animal life. Our inability to increase the number of whooping cranes shows how difficult it is to regenerate some untamed species once they have been depleted.

ANIMAL-MAN INTERACTION. The symbiotic relationship between man and animals has brought striking changes in both. As man studies his animals he learns more about himself—sociologically, physiologically, mentally, emotionally, and economically. At an early date animals became a part of man's religion and of his recorded art. The behavioral patterns of animal societies give insights into our whole ecosystem which are deserving of the increasing studies being directed toward them. Man still has much to learn from the living things over which divine edict has given him dominion. Courtship, sex, love, compassion, devotion, aggression, sharing, group defense, leadership, and many other manifestations in animal societies contain valuable lessons for the human race.

Scientifically and economically the contribution of domestic animals has already been virtually beyond measure. Animals have served as a mirror in which man has scrutinized his own physical structure and function, lengthening his life expectancy through improved knowledge of reproduction, nutrition, medicine, and surgery. The economic advantages enjoyed by those areas of the world with an advanced animal agriculture need not be documented here. Cultural and technological advances, including man's outreach into space, likewise have been centered principally in those countries with the most advanced animal agriculture.

ANIMAL CLASSIFICATION AND IDENTIFICATION

It is possible to know many breeds of animals without being an expert in animal production. Such knowledge, however, can only be acquired through the study of specific facts about breeds and some practice in careful and systematic observation of animals.

Since domesticated animals comprise only a small part of all animal life in the world, it is best to start the study by observing where they fit into the overall scientific classification. All living things are classified under one of two major groups called kingdoms—animal and plant. Each kingdom is subdivided so that every plant or animal can be specifically identified. In the animal kingdom the major subdivisions are phylum, class, order, family or suborder, genus, and species. Animals comprising a species are closely related and interbreed freely. Most farm breeds of a specific kind of livestock such as cattle, hogs, sheep, and horses have originated from one species within the genus (cattle excepted). Different species within a genus are related in general but may or may not interbreed freely. Crosses at the genus level (such as goat and sheep) are usually not possible.

Table 1.1 gives a scientific classification of animals; all breeds discussed in this book can be readily classified by using this table.

The animal kingdom includes all animals collectively. Animals in the phylum Chordata include only those with a backbone or rudiment of a backbone. The class and order further narrow the group significantly. In speaking of animals by their scientific name, only the genus and species terminology are used; thus *Bos taurus* indicates a large group in the cattle family, and *Sus scrofa* indicates a large group in the swine family.

The information needed to determine that the genus is *Bos* and not *Equus* is not great, but it is important as the first step in knowing breeds. Slightly more knowledge is needed to separate certain other genera, e.g. *Capra* (goats) from *Ovis* (sheep). However, careful observation of a few animals of each family makes such identification possible.

BREED IDENTIFICATION. A definite procedure in observing animals will develop good habits of observation and prevent neglect of minor but important characteristics in the process of breed identification. It is suggested that one observe first the overall body characteristics of an animal. This indicates the breed's major use. Following the use determination, the observation of color and markings, head and ear shape, and special distinguishing characteristics are in order. General facts about breeds plus some detailed observation enables positive identification of most breeds now existing in the United States. It is best to study several animals of a breed before making a final decision; any one animal may be rather far from the average or standard of the population.

The reason for observing animals in this order is that breeds are built first for specific purposes and second for discernible traits which characterize breeds and make them unique. Since general body conformation is the most indicative character of animal usefulness, it is proper to study it first. Thus if the purpose of a horse is draft and not racing, the large massive body with much muscling and substance indicates that the breed must be one in the draft classification. Once this is ascertained, the problem of identification is limited to only five breeds in the United States.

Producers of breeds have been proud of their livestock and have often limited their selection and registration to fixed or specific color and markings; thus color is usually the second character to observe and

TABLE 1.1. Taxonomic Classification of Animal Kingdom

Animal Kingdom

Phyla: Twenty-three phyla are generally recognized. These include many of positive or negative economic importance in relation to domesticated animals, because of their role as parasites, disease carriers, etc. The phylum of domesticated animals is
Chordata

    Subphyla: Four of these, including
        Vertebrata

    Classes: Marsipobranchii (lamphrey)
        Selachii (shark, dogfish)
        Bradyodonti (rabbit-fish)
        Pisces (bony fish)
        Amphibia (salamander, frog, toad)
        Reptilia (turtle, alligator, lizard, snake)
        Aves (birds)

        Orders: Struthioniformes (ostrich)
            Rheiformes (rhea)
            Casuariiformes (cassowary, emu)
            Apterygiformes (kiwi)
            Tinamiformes (tinamou)
            Gaviiformes (diver)
            Podicipediformes (grebe)
            Sphenisciformes (penguin)
            Procellariiformes (shearwater, diving petrel)
            Pelecaniformes (pelican, frigate bird)
            Ciconiiformes (stork, spoonbill, heron)
            Phoenicopteriformes (flamingo)
            Anseriformes (screamers, ducks, geese, swans)
            Falconiformes (vultures, ospreys, hawks)
            Galliformes (pheasants, turkeys, chickens, guinea fowls)
            Gruiformes (bitterns)
            Charadriiformes (plovers, gulls)
            Columbiformes (pigeons, sand-grouse)
            Psittaciformes (parrots, budgerigars)
            Cuculiformes (cuckoos, road-runners)
            Strigiformes (owls)
            Caprimulgiformes (oil birds, nightjars)
            Apodiformes (swift, hummingbird)
            Coliiformes (mouse bird)
            Trogoniformes (trogon)
            Coraciiformes (kingfisher, roller)
            Piciformes (woodpecker, toucan)
            Passeriformes: Several suborders which include songbird, crow, starling, raven, swallow
      Mammalia
        Subclasses:
        Prototheria
          Order: Monotremata (duckbill, spiny anteater)
        Theria
          Orders: Marsupialia (opossum, pouched mouse, wombat, koala, kangaroo)
              Insectivora (hedgehog, shrew, mole)
              Dermoptera (flying lemur)
              Chiroptera (bat)
              Primates (man, monkey, chimpanzee, gorilla)
              Edentata (three-toed sloth, armadillo, giant & lesser anteaters)
              Pholidota (pangolin, scaly anteater)
              Lagomorpha (rabbit, hare, cottontail)

TABLE 1.1. *(continued)*

Mammalia *(continued)*
    Rodentia: Several suborders which include squirrel,
      woodchuck, beaver, hamster, rat, mouse, guinea
      pig, chinchilla
    Cetacea (whale, dolphin, porpoise)
    Carnivora
      Suborder: Fissipeda
        Genus: Canis (wolf, dog, jackal)
              Vulpes (fox)
              Ursus (bear)
              Thalarctos (polar bear)
              Procyon (raccoon)
              Mustela (ferret, weasel, mink, ermine)
              Martes (marten, sable)
              Meles (badger)
              Taxidea (American badger)
              Felis (cat)
              Several others
    Pinnipedia (sea lion, seal, walrus)
    Tubulidentata (aardvark)
    Proboscidea (African & Asian elephants)
    Hydracoidea (tree hydrax, coney)
    Sirenia (dudong, manatee)
    Perissodactyla (uneven-toed)
      Suborder: Hippomorpha
        Genus: Equus (horse, zebra, donkey)
      Suborder: Ceratomorpha
        Genus: Tapirus (tapir)
              Rhinoceros
    Artiodactyla (even-toed)
      Suborder: Suiformes
        Genus: Sus (pig)
              Tayassu (peccary)
              Hippopotamus
      Suborder: Tylopoda
        Genus: Lama (llama, alpaca, vicuna, guanaco)
              Camelus (camel, dromedary)
      Suborder: Ruminantia (true ruminants)
        Genus: Tragulas (chevrotain)
              Moschus (musk deer)
              Dama (fallow deer)
              Cervus (red deer, elk, wapiti)
              Alces (moose)
              Rangifer (reindeer, caribou)
              Okapia (okapi)
              Giraffa (giraffe)
              Taurotragus (eland)
              Bubalus (buffalo)
              Bos (cattle\
                B. *taurus* (European cattle)
                B. *indicus* (Indian humpback cattle)
              Bison (American buffalo or bison)
              Hippotragus (roan antelope)
              Antilope (Indian antelope)
              Gazella (gazelle)
              Rupicapra (chamois)
              Ovibos (musk ox)
              Capra (goat)
              Ovis (sheep)
              Poëphagus (yak)

Adapted from Rothschild (1961), *A Classification of Living Animals.*

will eliminate many of the remaining breeds. If the breed has not been identified in the first two steps, a study of head and ears is necessary. Special distinguishing markings are good for checking conclusions reached from preliminary study.

To some, such a routine procedure may seem monotonous. However, breed identification is an accurate study, and following a correct procedure assures the recognition of a maximum number of breeds with a minimum of effort. It is what the breed can do that makes its identification desirable. Probably the most important fact to remember is that breeds are the genetic banks for all livestock producers and are truly the livestock man's "goose that lays the golden egg." Breeds hold the wealth of centuries of man's labors involved in (1) domestication of animals, (2) production of animals for specific purposes, (3) breed establishment and development, and (4) production breeding for efficiency and quality of product. They must be continued and further improved; thus they are worth knowing as an important part of the world's storehouse of usable knowledge.

FUTURE OF BREEDS. Man has three opportunities to improve the germ plasm pool for producing animal products.

1. Domesticate more wild animals.
2. Distribute more generally some breeds that are not now used widely.
3. Improve the breeds that are now being used heavily.

Domestication of animals native to arctic and antarctic regions appears to have much to offer. Cataloging all breeds of the world for production merit and then testing each in areas of most promise should be even more profitable. As pointed out throughout the book, improvement within breeds holds immense possibilities.

The most important pressure for germ plasm improvement for tomorrow's production is the world "population explosion." Table 1.2

TABLE 1.2.   Relative Animal Efficiency in Converting Animal Feed Nutrients to Human Food

Animal Product	Energy (%)	Protein (%)	Gross Edible Product Output per Unit of Feed Intake (%)
Milk	20	30	90
Beef	8	15	10
Lamb	6	10	7
Pork	15	20	30
Eggs	15	20	33
Chicken (broilers)	10	25	45
Turkey	10	20	29

Source: 6th FAO Report, Rome, 1968.

shows the relative efficiency of converting feed to animal products for human consumption. The rapid population increase of this century can mean one of two things: rapid increase in production merit of our domesticated animals or decreased consumption of animal products on a per capita basis. The preference for a choice steak over a peanut butter sandwich is not expected to diminish among American consumers; but the continuing opportunity for them to exercise this option will require increasingly efficient production of superior products by our commercial livestock.

DEVELOPMENT AND USE OF BREEDS

During the thousands of years that our farm animals have been bred by man since first domesticated, they have changed in temperament, form, fertility, color, hair coat, size, and usefulness. Changes made within any one period were small and not always in the same direction. Isolation of some animals by land and water barriers plus unusual selection ideals of some livestock owners resulted in some breeding groups differing significantly from the average of the species. Man recognized the difference and usefulness of these distinct types and continued to produce them and to form new ones. But it was not until the middle of the 18th century that a start was made in the formation of what we now call purebred breeds.

A breed of livestock as defined by animal scientists today is a population or group of animals that has a common ancestry and possesses certain markings and other characteristics which distinguish the group from other animals within the species. Man's ability to produce widely differing types within a species is perhaps best illustrated by the many distinct breeds of dogs. Breeds are generally sponsored by cooperative breeder associations which in most cases record the ancestry of animals that are eligible for registration as purebreds. The associations also aid breeders in continuous development, promotion, and sale of breeding stock.

ESTABLISHMENT OF BREEDING PRINCIPLES. Robert Bakewell of England (1725–95) is credited with establishing the pattern of animal breeding that has led to modern animal improvement and the formation of breeds. He lived during the period of the Industrial Revolution of England and was quick to see the opportunities as well as the needs being created for agriculture with the removal of many people from the country to towns and cities. He also recognized that the farm animals then being produced were ill suited for mass production of animal products, and he undertook to improve them, both by breeding and feeding. Although he is known chiefly for his breeding work, records show that he was also prominent in introducing to England turnips and other root crops for livestock feeding.

Bakewell first became known as an outstanding livestock breeder and producer when he began managing his estate at Dishley in 1770. At that time he was developing Shire horses, Longhorn cattle, and Leicester sheep. His breeding program included selection for improvement in (1) utility of form, (2) fleshiness or muscling, and (3) ability to gain in the feedlot. He was extremely successful in producing animals that were superior to those of his neighbors, and his stock was soon in great demand. Many also sought his advice on livestock improvement.

Bakewell told others little about his methods and wrote less. However, many young stockmen came to work for him, and they studied his breeding principles: (1) Select for a type based on usefulness; (2) breed the best to the best; (3) like tends to beget like; (4) inbreed to fix characters and bring refinement. In selecting the best, Bakewell was insistent on considering only needed characteristics. His favorite slogan in cattle breeding was "All is useless that is not beef." It is said that he often pickled bones or parts of animals for comparison in order to be certain that he was making improvement in the characteristics he wanted. Many of Bakewell's students were also successful in improving their livestock. The Colling brothers, who founded the Shorthorn breed that has spread around the world, once worked with Bakewell.

The success of Bakewell and his associates soon created a demand for all improved stock of the area. Orders came from other parts of England and many foreign countries. Prices for this improved stock were high and encouraged further improvement. In the beginning the breeders had not planned to develop breeds as we know them today; they only wanted superior animals for use on their own farms and for sale to other producers. But since the improved stock was similar in many ways to the unimproved stock of the area, it soon became evident that breeders would have to do something to protect themselves and prospective buyers from unscrupulous producers who claimed they also had good stock. Breeders had been privately recording the ancestry of their stock but were not releasing it for fear they would give away their breeding secrets. The threat of inferior animals taking part of the newly developed market and destroying the reputation of improved groups of animals convinced breeders that secrecy was a much greater hazard to their business than was the record of ancestry of their animals. In fact it was soon evident that only the latter could save them. Thus identification of groups as breeds, publication of pedigrees, and formation of associations for promotion of breeds became the standard procedure for all improved animal groups.

BREEDS AND BREEDER ASSOCIATIONS. The typical history of the formation of early breeds of livestock was about as follows. First, there existed a group of animals which were more useful than the average for specific purposes but not distinctly different in pedigree from others of the area. Second, selections were made of the best animals of such a group by a few outstanding breeders who culled heavily against undesirable characteristics and poor production, and these were inbred to better fix the heredity of the best traits. Third, the merit of the improved animals was recognized by other livestock men, and breeders were forced to start a central herd or flock book for registration of improved animals. Fourth, an association was established for the purpose of safeguarding the purity of the breed, maintaining a record of ancestry, developing continuous improvement, and promoting the use and sale of breeding stock.

In recent years many new breeds have been formed by private breeders and by specialists at research institutions. These breeders have followed the same basic steps established by Bakewell and his associates. Step one has usually been slightly altered in that it has consisted of crossing superior individuals of two or more breeds and using this population as foundation stock instead of starting with a group of animals already in existence. Step two has usually proceeded with more rapid inbreeding than was done by founders of early breeds.

Many new breeds will undoubtedly be established in the future. Methods of developing these will probably be similar to those used now and in the past. Some of the new breeds will fail. Those that have merit and are sponsored by hard-working intelligent breeders will find their place in tomorrow's commercial livestock production programs.

The first herdbooks often preceded the establishment of breeds and have done much to set the standard for registration of today's breeds. The first published herdbook on record was *An Introduction to the General Stud Book,* which was established in 1791 and recorded pedigrees of Thoroughbred horses that had won important races. It did not publish the pedigrees of all members of the Thoroughbred breed and thus was incomplete as a registry book. The identification and publication of pedigrees of outstanding producers, however, is the principle of advanced registry used by many associations today to call attention of breeders to outstanding individuals.

The second herdbook to be published was the *Coates Herdbook,* established in 1822 by George Coates as a private enterprise to serve the breeders of Shorthorn cattle. It was started long before the Shorthorn breed was officially established. Some of the early breeders objected to furnishing pedigrees and other data to Coates, but it soon became evident that knowledge of pedigrees was necessary to protect breeders and buyers and no one could remember such information for an entire breed. Data published in the early herdbooks generally included (1) number and name of the animal, (2) date of birth, (3) ancestry, (4) breeder, and (5) color markings. Many present-day breed associations publish production data on registration certificates or in supplemental publications.

Qualifications for entry of an animal to a breed generally include (1) being an offspring of registered parents and (2) freedom from defects. Some breeds require certain minimum production standards. A few breeds maintain open books and allow animals to enter if they are the offspring of several successive crosses of registered sires of the breed and have high production merit. Other breeds have closed herdbooks; no individual can be recorded unless both parents are registered. Livestock breeders often use the term *purebred* to mean any animal whose parents are registered with a breeders' association group. The term *registered* is always used to mean an animal that is registered in the association's herdbook, having met all the requirements for registration. However, the term purebred may be used to mean the same as registered.

USE OF BREEDS. In North America and much of the rest of the world, purebred breeds are the source of germ plasm for all commercial production. Breeds have real and lasting merit for this purpose. Their genetic makeup is more predictable than that of animals from random breeding groups, giving producers a somewhat constant supply of similar germ plasm for use in any breeding system desired. Thus commercial producers may use (1) purebreeding, (2) grading up to some breed through continuous use of purebred sires of that breed, and (3) crossbreeding. Without purebred breeds such systems would be impossible for commercial producers.

The practice of selecting within closed populations such as breeds also serves to hold merit characteristics that are not easily measured and/ or highly hereditary. Many desirable qualities of farm animals are of this nature.

In the past some producers have had a tendency to feel that purebreds were the end goal. This has resulted in selection for many fancy points instead of commercial production merit. If this is carried very far, the breed soon becomes a liability rather than an asset to scientific animal agriculture. Fortunately, modern breeders of purebreds are selecting more and more for production merit, thus making breeds continually more valuable to mankind. No breed is perfect. No breed is pure for its inheritance. Thus breeds are ever changing and do not need to become outmoded if we understand them and mold them to meet our needs.

READING LIST

Anderson, A. L., and J. J. Kiser. 1963. *Introductory Animal Science.* Macmillan, New York.
Briggs, H. M. 1969. *Modern Breeds of Livestock.* Macmillan, New York.

Campbell, J. R., and J. F. Lasley. 1969. *The Science of Animals That Serve Mankind*. McGraw-Hill, New York.

Economic Research Service, USDA. *Dairy Situation*. U.S. Government Printing Office, Washington, D.C.

Economic Research Service, USDA. *Livestock and Meat Situation*. U.S. Government Printing Office, Washington, D.C.

Economic Research Service, USDA. *Poultry and Egg Situation*. U.S. Government Printing Office, Washington, D.C.

Hodgson, R. E. (ed.). 1961. *Germ Plasm Resources*. Publ. No. 66, American Association for the Advancement of Science, Washington, D.C.

Koch, Charles R. 1968. Safe entry for new genes. *Farm Quarterly*, Fall, p. 42.

Phillips, R. W. 1948. *Breeding Livestock Adapted to Unfavorable Environments*. FAO Agricultural Studies No. 1, Washington, D.C.

USDA Yearbook of Agriculture, 1936. U.S. Government Printing Office, Washington, D.C.

USDA Yearbook of Agriculture, 1937. U.S. Government Printing Office, Washington, D.C.

[CHAPTER TWO] Cattle [BOVINES]

THE BOVINE—idolized, eulogized, and sometimes denigrated in the art, music, religion, literature, and drama of man—is more closely interwoven with the fabric of man's culture than any other animal. *Bos taurus* inhabits the globe from the fringes of the tropics to the approaches of the polar ice caps, and *Bos indicus* encircles the earth's tropical mid-section. Considerable overlapping and intermixing of the two species have occurred. The versatile cow is truly "the foster mother of the human race"—as provider of milk and meat; as a draft animal or beast of burden; as a source of hides for clothing, shelter, and equipment; as a supplier of manure for fuel and fertilizer; and as an object of religion and medium of exchange.

Four groups of cattle are generally recognized within the Bovidae family, suborder Ruminantia, order Artiodactyla. The taurine group includes *Bos taurus* (ordinary cattle) and *Bos indicus* (humped cattle). These are the two bovine species of major current importance in the Western Hemisphere and rather generally throughout the world. The bibovine group includes the gaur *(Bibos gaurus)*, the gayal *(Bibos frontalis)* and the banteng *(Bibos sondaicus)*. The bisontine group includes the European bison *(Bison bonasus)* and the American bison or buffalo *(Bison bison)*, which will also interbreed with *Bos taurus*. Sterility or limited fertility, sometimes sex related, occurs in these interspecies hybrids. The bubaline group includes the African buffalo *(Bubalus caffer)* and Indian buffalo *(Bubalus bubalis)*, frequently referred to in current literature as Murrah buffalo or water buffalo, which are important food producers and draft animals in their areas of distribution.

Evidence suggests that *Bos taurus* evolved from *Bos primigenius* (the great ox [aurochs] of Europe mentioned in Caesar's writings) and *Bos longifrons* (a smaller type found on the British Isles prior to the introduction of aurochs by the Anglo-Saxons). *Bos indicus* probably descended from the wild Malayan banteng.

NUMBERS AND DISTRIBUTION

There are well over 1 billion head of domesticated bovines in the world, of which about 10% are water buffalo. The major cattle-raising countries are (1) India, (2) United States, (3) USSR, (4) Brazil, (5) China,

[16]

(6) Argentina, (7) Pakistan, (8) Mexico, (9) Ethiopia, and (10) France.

On January 1, 1970, there were 112,330,000 head of cattle and calves in the United States, of which 37,433,000 were identified as beef cows 2 years old and over and 13,875,000 as milk cows 2 years old and over. During the quarter-century since World War II there has been a steady decline in the number of milk cows in North America, with little change in total milk production, and a concurrent increase in the number of beef cattle. These changes in opposite directions have been of about equal magnitude, resulting in little change in total cow numbers. In terms of producing-age females (cows 2 yr. old and over), the leading beef cow states are (1) Texas, (2) Oklahoma, (3) Missouri, (4) Nebraska, (5) Kansas, (6) South Dakota, (7) Montana, (8) Iowa, (9) Mississippi, and (10) Kentucky; the leading dairy cow states are (1) Wisconsin, (2) New York, (3) Minnesota, (4) California, (5) Pennsylvania, (6) Iowa, (7) Michigan, (8) Ohio, (9) Missouri, and (10) Kentucky.

INTRODUCTION AND DEVELOPMENT

Vast herds of American bison roamed the plains of North America, providing sustenance for the plains Indians prior to the westward movement of the first white settlers. The white men exploited these herds for their hides almost to the point of extinction before laws were passed to preserve the species. Hybrids of bison and cattle crosses, often referred to as "cattalo," have been produced and studied.

The first taurine cattle to arrive in the Americas were of Spanish origin, brought by Spanish explorers in the 1500s. These hardy, long-horned cattle thrived and multiplied in a semiwild state throughout southwestern United States and Mexico and southward into Central and South America, where they became the progenitors of the Longhorns of North America and the Criollo cattle of the Spanish-speaking countries to the south. These rangy, light-muscled cattle were not particularly

FIG. 2.1. *Longhorn 3-year-old cow with her first calf.* (Bureau of Sport Fisheries and Wildlife, U.S. Dept. of Interior)

noteworthy for the production of either meat or milk. While the Criollo-type Blanco Orejinegro breed is still found in small numbers in South America, the main use of these Spanish cattle has been for grading up with British or Zebu breeds.

As English settlers established colonies along the East Coast, cattle of dual-purpose type were brought with them from England. Many of the earliest ones, brought in the early 1600s, were undoubtedly Durham (Milking Shorthorn) or Devon types well suited at that time for production of milk and meat and for use as draft oxen. Many cattle of this type moved west with the homesteaders. Specialized beef and dairy breeds from the British Isles and the European continent began arriving in substantial numbers in the early 1800s and continued until after the turn of the century. Most of the cattle breed registry organizations in the United States were formed during the mid to late 1800s, frequently predating comparable herdbook societies in the countries of their origin.

The earliest importations of recognizable breed types do not always coincide with those to which present-day animals are traceable. The very early Durham types have already been noted. It is also known that black and white Holstein-type cattle were brought into the Mohawk Valley of New York State during the 1621–25 period by Dutch settlers, whereas the earliest cattle of this type known to contribute to today's Holstein breed were imported by W. W. Chenery of Massachusetts in 1850. Shorthorn cattle were imported to Virginia as early as 1783, but records as to the disposition of these cattle were not kept. Thomas Bates, a noted breeder in England, developed some of the famous early strains of dual-purpose type Shorthorns. These Bates Shorthorns rode a crest of popularity in the United States starting about 1850, with special emphasis on the highly inbred Duchess family. Inflated prices prevailed for these cattle for about 15 years until "Scotch" Shorthorns of straight beef-type breeding began to attract increasing attention along with the Angus and Hereford breeds.

The British Isles have been the major source of improved breeds for the establishment of the livestock industry throughout the Western Hemisphere. Once the large-scale introductions of cattle breeds got under way in the United States, Canada, Mexico, Argentina, Uruguay, and other major cattle-producing areas in the temperate zone, a clear demarcation line developed between cattle kept for beef and those kept for milk. Dual-purpose cattle continued to hold a place on many general farms until about 1950, but in general the beef-producing and milk-producing segments of the cattle industry followed sharply contrasting routes in their choice of breeds and selection emphasis within those breeds. The European continent, on the other hand, has traditionally emphasized the dual-purpose role of most cattle breeds, even with those such as the Holstein-Friesian and Brown Swiss which founded specialized dairy breeds in North America. It should be recognized that, since all

THE CHANGING PROFILE OF TYPE

BEEF CATTLE DAIRY CATTLE

FIG. 2.2. *Young Longhorn steers. Earliest type in the Western Hemisphere.*

FIG. 2.5. *European type Friesian cow.*

FIG. 2.3. *Hereford show steer. About 1950.* (American Hereford Assn.)

FIG. 2.6. *Holstein-Friesian show cow. Mid-20th century.*

FIG. 2.7. *Modern type Holstein-Friesian cow. All-time All-American winner.*

FIG. 2.4. *Hereford steer. 1970 ideal concept.* (American Hereford Assn.)

cattle produce beef, the real distinction between dairy or dual-purpose cattle and beef cattle is that the latter have no merit as milk producers beyond that required for their calves. Under some selection programs, even this minimal milk-producing capability has often been marginal in the specialized beef breeds.

The distribution of cattle breeds in North America can be characterized briefly. Herefords became the predominant breed on the western range and in much of the hilly Appalachian area. Angus and Shorthorns found favor on the more productive farms of the Corn Belt and the East. Brahmans became important, principally for crossing with British breeds, along the Gulf Coast from Florida to Texas. The first American-developed cattle breed, the Santa Gertrudis, was developed on the King Ranch in Texas by blending the Brahman and Shorthorn breeds. There are actually many recognized breeds or specialized types of Zebu (Indian hump-back cattle), some of which have been selected for milk production. The term "Brahman" was coined in America to apply collectively to the meaty types of Zebu that were first introduced, representing such breeds as Nellore, Ongole, Guzerat, and Gir. Similar breeds (collectively called Cebú) underlie the large beef-producing herds in Brazil, Colombia, and Venezuela. Zebu breeds which have been selected for dairy merit include the Sahiwal, Red Sindhi, and Tharparkar.

Specialized dairy breeds became established near metropolitan centers throughout the country and in major areas of butter and cheese production settled by Dutch and Scandinavian people with dairying traditions. Wisconsin, Minnesota, and New York are good examples of this latter type of concentration associated with ethnic groups.

EVALUATING AND IMPROVING PERFORMANCE

Utility underlies the development and distribution of all improved types and breeds of livestock. Generally speaking, the following sequence has obtained: (1) development of a useful and distinctive type, (2) recognition of merit, (3) merchandising and distribution, (4) improvement in the direction of established ideals.

Beef production merit was recognized by Charles Darwin in *The Variation of Animals and Plants under Domestication* (1897): "Not only should our animals be examined with the greatest care whilst alive, but as Anderson remarks, 'their carcasses should be scrutinized so as to breed from the descendants of such only as, in the language of the butcher, cut up well.'" Similar early evidence of the significance attached to productive traits can be found in letters and other writings quoting the milk or butter yields of improved dairy cattle.

In retrospect one can observe many periods of unwarranted emphasis on traits contributing little or actually retarding progress in utilitarian values. Color fads, fancy points of show-ring conformation, excessive pedigree emphasis, and other extremes can be cited. But utilitarian ideals also change with time and local circumstances. In total, progress has been made, and the genetic resources necessary to adjust for unwise or changing ideals have always been retained somewhere within the species.

Performance testing has been widely employed in the improvement of dairy cattle since early days. The objective measurement of economic traits in beef cattle is a more recent development, which began to accelerate rapidly about 1960. Both central testing stations and on-the-farm testing are employed in testing cattle for dairy or beef merit. Independent breeder-sponsored or breed association–sponsored milk and butter tests, covering short periods of usually one week or more, were used as a promotion and merchandising aid by early importers and breeders of the specialized dairy breeds. Following the development in 1890 of the Babcock test for milk fat by S. M. Babcock of Wisconsin, fat content and volume of milk became the official criteria of evaluation. Selective testing of individual cows for part or complete lactations was initiated by the Holstein association's official Advanced Register program in 1895, with the other dairy and dual-purpose breeds following suit with comparable programs.

Denmark gave birth in 1895 to the complete-herd concept of production testing which has been copied by all major dairy areas of the world. Danish immigrant Helmer Rabild introduced this Cow Testing Association idea to the United States in 1905 in Newaygo County, Michigan. The merit of continuous testing of every cow in the herd under the monthly sampling supervision of an official tester was quickly recognized by the USDA, and Rabild was hired to help establish this system among U.S. dairymen. From this beginning the Dairy Herd Improvement Association (DHIA) evolved as a part of the Cooperative Federal-State Extension program to eventually become today's basic testing program for all dairy cattle and dairy goats, both registered purebreds and otherwise.

Official on-the-farm records of weaning weights and other beef cattle traits, which got under way about a half-century later under the impetus of state beef improvement association leadership, is an adaptation of the DHIA concept. Credit for originating the central testing station system of individual or progeny testing of dairy cattle, beef cattle, and swine also goes to Denmark.

SIRE EVALUATION AND ARTIFICIAL INSEMINATION. Emphasis on production testing soon brought to the attention of dairymen the vast differences in the genetic merit of sires. A USDA germ plasm survey in the 1930s further focused attention on the importance of sire evaluation and selection and culminated in the establishment of routinely

computed USDA sire proofs based on a comparison of DHIA production
of a bull's daughters with that of their dams.

The advent of artificial insemination (AI) about 1940 as a feasible
routine technique with cattle motivated additional investigations toward
more accurate sire evaluation methods. The long-time storage and un-
limited selection possibilities subsequently opened up by the British development of semen freezing added urgency to the search for more adequate sire comparison criteria. In 1962 the USDA accepted the method
developed by Cornell University of comparing a bull's daughters with
their herdmates to replace the daughter-dam proofs. A *herdmate* in
the context used here is another cow of the same breed, in the same herd,
that started her lactation in the same year and season as the bull's
daughter, excluding paternal sisters. Canada uses a similar evaluation
procedure, except that contemporaries of the daughters are used rather
than using records of cows of all ages adjusted to a mature basis as
under the U.S. system. Refinements continue in this dynamic endeavor
to improve genetic merit for productive traits. Increasing selectivity by
individual dairymen, due to virtually worldwide frozen semen availa-
bility of any sire, brings into the picture the possibility of direct within-
herd contemporary comparisons among daughters of different sires used
in such herds. The future will bring many changes.

An all-breed nationwide beef cattle performance evaluation program available to cattle of any genetic background is Performance
Registry International, organized in 1955. It awarded its first Golden
Certified Meat Sire certificate in 1962 to the Angus bull Emulous 7000.
Official recording of rate and composition of gain by a sire progeny group
provides the basis for certification of sires, indicating their ability to
transmit performance and carcass traits. As of January 1, 1971, a total of
201 sires, representing all the established beef breeds, had been so recognized. Traits considered in the progeny evaluation are age at slaughter,
carcass weight, rib-eye area, fat thickness, and marbling score. The
paucity of performance testing which so long characterized beef cattle
breeding and merchandising is rapidly being corrected by an awakened
industry. The traditional tight restrictions on AI use, imposed by
registry societies of all except those of the dairy breeds, will gradually be
relaxed as performance testing singles out the sires worthy of extensive
use. Commercial interests will demand it, and such a trend is already in
evidence among the beef breed associations. The impact of performance
superiority through extensive AI use has already been demonstrated
in dairy cattle by sires with 10,000 or more production-tested daughters
averaging as much as 1,000 lb. milk or more above herdmates. Frozen
semen from a wide choice of herd-improving sires can be purchased for
$5–10 per ampule. The extreme price for an ampule of scarce semen
from an especially popular sire no longer producing has run into five
figures.

The arrival in the 1960s of the so-called "exotics" or "new breeds" for use in the beef cattle industry is closely tied to the AI industry. A strict quarantine facility and procedures developed by the Canadian government paved the way for these introductions from Europe. Their maximum use is being achieved by frozen semen distribution throughout the North American continent. Grading-up registry procedures are a necessity for these breeds, in contrast to the traditional closed herdbook concept adhered to by most breeds introduced to North America heretofore. Even with the established breeds, AI has had a profound influence on registry policies. Most dairy breed registry organizations have recently adopted or have under consideration procedures for accepting into an open herdbook animals identifiable with the breed but lacking a complete record of registered ancestors.

SOME PERFORMANCE BENCH MARKS. Performance is ever changing and records are continually being broken. It may be informative, however, to note some production highlights as of this writing. Some indications of average performance characteristics are given in the breed summary (Table 2.3) at the end of the chapter.

In dairy production circles, annual herd averages of over 20,000 lb. milk and 700 lb. fat are increasingly being achieved. Individual lactation records of 305 days are routinely used for comparisons, adjusted to a mature equivalent 2x milking basis. In most breeds individual records up to 365 days in length are officially recognized, and many of the earlier records were made on 3x milking. Table 2.1 shows the top individual lactation performances by breeds as of 1971.

TABLE 2.1. Top Lactation Performances as of 1971

Breed	Year	Name and Place	Age (Yr.–Mo.)	Length of Record (days)	Times Milked Daily	Milk (lb.)	Fat (lb.)
Ayrshire*	1970	Fairdale Betty Gem (Mass.)	6–7	305	2	32,250	1,109
	1963	Bob's Pansy Girl (Maine)	5–11	305	2	20,040	1,213
Brown Swiss	1958	Lee's Hill Keeper's Raven (N.J.)	9–9	365	3	34,850	1,570
	1959	Letha Irene Pride (N.J.)	11–5	365	3	34,810	1,733
Guernsey	1970	Fox Run A.F.C. Faye (Tenn.)	5–4	365	2	32,110	1,397
	1971	Fox Run A.F.C. Faye (Tenn.)	6–6	365	2	31,040	1,736

* Only 305-day records recognized.

TABLE 2.1. (continued)

Breed	Year	Name and Place	Age (Yr.–Mo.)	Length of Record (days)	Times Milked Daily	Milk (lb.)	Fat (lb.)
Holstein	1971	Skagvale Graceful Hattie (Wash.)	7–3	365	2	44,019	1,505
	1960	Princess Breezewood R.A. Patsy (Ohio)	5–2	365	2	36,821	1,866
Jersey	1970	Trademarks Sable Fashion (Tex.)	8–0	365	2	29,320	1,550
Milking Shorthorn	1962	Hazelbrook Red Jane 8th (Pa.)	5–3	365	2	23,735	895
	1926	Ruth B. (Wis.)	8–5	365	3	21,641	957

The lack of a long-time nationwide program for recording authenticated individual performance data for beef cattle precludes the listing of top performers. More criteria enter into the evaluation of beef production merit than of dairy merit, and only recently have these been set forth and quantified by actual measurements. Following goals similar to Performance Registry International, the Beef Improvement Federation (BIF) was established in 1968 as a cooperative effort of extension, research, and the cattle industry to set up uniform guidelines for beef cattle performance evaluation. These guidelines encompass preweaning and postweaning performance measurements and carcass evaluation. With regard to some important performance traits it can be noted that adjusted 205-day weaning weights of 600 lb. or more, feedlot gains of 3 lb. per day or more, and 365-day weights of 1,000 lb. are increasingly being achieved, along with high-quality carcasses yielding 50–55% of trimmed retail cuts. The listing of recorded performance traits in beef cattle pedigrees is being used increasingly in merchandising, as has been routine practice for years in dairy cattle merchandising. Some performance data from bull testing station results are given in Table 2.2.

TABLE 2.2. Performance Data on Bulls in 140-day Test at Iowa Beef Improvement Association Test Station, 1970

Breed	205-day Weaning Wt. (lb.)	Test Period Av. Daily Gain (lb.)	365-day Wt. per Day of Age (lb.)
Angus	522	2.73	2.53
Brown Swiss	516	2.91	2.53
Charolais	599	3.24	2.95
Hereford	501	2.98	2.61
Polled Hereford	534	2.80	2.60
Red Angus	557	3.01	2.62
Shorthorn	505	3.08	2.73
Crossbred bulls (Charolais, Limousin & Simmental crossed with British breeds)	479	3.59	2.80

VISUAL APPRAISAL. A brief discussion of this time-honored and vastly overemphasized method of livestock evaluation is presented last. Relegating it to this position is more indicative of its relative merit than of its actual impact over the years.

The generally deleterious effect of the show-ring on the economic merit of all types of livestock has been due largely to (1) failure to recognize the show-ring as a strictly merchandising rather than an evaluation tool, and (2) overemphasis of many uneconomic traits or fancy points and use of artificial methods of achieving them. In meat animals undue emphasis on form and fatness rather than growthiness and muscling has been a major shortcoming. Compounding the error in beef cattle shows has been the use of dairy nurse-cows for fitting young show cattle. When an Angus or Hereford yearling bull said goodbye to his Holstein foster mother at the culmination of a successful show season, he was ill equipped by virtue of conditioning or any indication of true merit to make his way in the "real world" as a sire of profitable calves for a western rancher. Fortunately, even if late, this practice had been ruled out at most beef shows by 1960. Even with the most sensible fitting procedures and ideals of type, a single show-winning animal or small group of animals provides scanty evidence of breeding merit. The only importance that should be ascribed to a show-ring winning—and quite significant to the industry—is that of attracting attention to a sound, performance-based breeding program which can withstand the scrutiny of a more adequate genetic evaluation than the show-ring can ever provide.

Fat also effectively covers up imperfections of body form in dairy cattle. Overconditioning during the growing period at the same time exerts an unfavorable effect on future reproductive and productive performance of breeding females of all species. Dairy cattle exhibitors traditionally smoothed up their show animals by putting on fat. About 1950 some major dairy exhibitors turned their backs on this practice and brought out their show strings looking like working dairy cattle. Since that time overconditioned dairy animals have virtually disappeared from the show-ring due to strong discrimination against them.

Official score cards or "standards of excellence" have been used as guides for visual appraisal for a long time in most cattle breeds. In some instances visual appraisal has been utilized as a prerequisite to registry in the herdbook. Each of the major dairy breeds (and some beef breeds such as the Angus) has an official on-the-farm classification program available to breeders. Each eligible animal in the herd is scored by an official classifier designated by the breed registry organization. To the extent that visual appraisal of important structural traits can contribute to the breeding of more profitable animals, this evaluation of all animals in a herd in working condition has considerable merit. Many variations of this kind of on-the-farm evaluation are provided by AI units, state extension services, independent consulting firms, and other cattle in-

DAIRY COW UNIFIED SCORE CARD

Copyrighted by The Purebred Dairy Cattle Association, 1943, Revised, and Copyrighted 1957
Approved — The American Dairy Science Association, 1957

	Perfect Score
Breed characteristics should be considered in the application of this score card	

Order of observation

1. GENERAL APPEARANCE — 30

(Attractive individuality with, feminity, vigor, stretch, scale, harmonious blending of all parts, and impressive style and carriage. All parts of a cow should be considered in evaluating a cow's general appearance) — 10

BREED CHARACTERISTICS — (see reverse side)
HEAD — clean cut, proportionate to body; broad muzzle with large, open nostrils; strong jaws; large, bright eyes; forehead, broad and moderately dished; bridge of nose straight; ears medium size and alertly carried
SHOULDER BLADES — set smoothly and tightly against the body — 10
BACK — straight and strong; loin, broad and nearly level
RUMP — long, wide and nearly level from **HOOK BONES** to **PIN BONES**; clean cut and free from patchiness; **THURLS**, high and wide apart; **TAIL HEAD**, set level with backline and free from coarseness; **TAIL**, slender
LEGS AND FEET — bone flat and strong, pasterns short and strong, hocks cleanly moulded. **FEET**, — 10
short, compact and well rounded with deep heel and level sole. **FORE LEGS**, medium in length, straight, wide apart, and squarely placed. **HIND LEGS**, nearly perpendicular from hock to pastern, from the side view, and straight from the rear view

2. DAIRY CHARACTER — 20

(Evidence of milking ability, angularity, and general openness, without weakness; freedom from coarseness, giving due regard to period of lactation) — 20

NECK — long, lean, and blending smoothly into shoulders; clean cut throat, dewlap, and brisket
WITHERS, sharp. **RIBS**, wide apart, rib bones wide, flat, and long. **FLANKS**, deep and refined. **THIGHS**, incurving to flat, and wide apart from the rear view, providing ample room for the udder and its rear attachment. **SKIN**, loose, and pliable

3. BODY CAPACITY — 20

(Relatively large in proportion to size of animal, providing ample capacity, strength, and vigor)

BARREL — strongly supported, long and deep; ribs highly and widely sprung; depth and width of barrel — 10
tending to increase toward rear
HEART GIRTH — large and deep, with well sprung fore ribs blending into the shoulders; full crops; — 10
full at elbows; wide chest floor

4. MAMMARY SYSTEM — 30

(A strongly attached, well balanced, capacious udder of fine texture indicating heavy production and a long period of usefulness)

UDDER — symmetrical, moderately long, wide and deep, strongly attached, showing moderate cleavage — 10
between halves, no quartering on sides; soft, pliable; and well collapsed after milking; quarters evenly balanced
FORE UDDER — moderate length, uniform width from front to rear and strongly attached — 6
REAR UDDER — high, wide, slightly rounded, fairly uniform width from top to floor, and strongly — 7
attached
TEATS — uniform size, of medium length and diameter, cylindrical, squarely placed under each quarter, — 5
plumb, and well spaced from side and rear views
MAMMARY VEINS — large, long, tortuous, branching — 2
"Because of the natural undeveloped mammary system in heifer calves and yearlings, less emphasis is placed on mammary system and more on general appearance, dairy character, and body capacity. A slight to serious discrimination applies to overdeveloped, fatty udders in heifer calves and yearlings."

Subscores are not used in breed type classification. — **TOTAL** — **100**

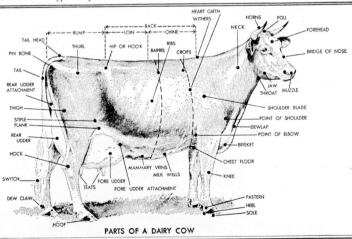

PARTS OF A DAIRY COW

dustry groups. Details on any of these programs can be obtained from the appropriate breed association or other source, usually well publicized through the breed magazines.

A LOOK AHEAD

The last 30 years of the current century promise to be fascinating ones in the cattle industry. Commercial interests are increasingly dictating the breeding programs and management practices in the industry as contrasted to the traditional dominance of the production scene by purebred interests. Such concepts as the "sacredness" of ancestral purity, color pattern consciousness, the dichotomy between dairy and beef cattle interests, and the precedence of form over function are giving way to emphasis on those factors which can best serve the interests of the consumer of animal products and the producers who supply them.

Open herdbooks are now a functional reality. Crossbreeding either as a formal breeding system or as the initial step in developing a genetic pool for synthesizing new breeds is no longer taboo, even to the extent of including dairy cattle genes as a part of the genetic resource for beef production. Establishment of the Red and White Dairy Cattle Association and subsequent establishment by the Holstein-Friesian association of a registry for purebred Holsteins exhibiting the recessive red color pattern are examples of more objective thinking than existed a decade ago. Animal health control facilities and AI use permitting the frantic introduction of new blood from Europe into the North American beef industry provide major speculation in future progress.

The race will go to the swift, and few prognosticators would even venture a guess as to the favorite at this point.

READING LIST

Briggs, H. M. 1969. *Modern Breeds of Livestock.* Macmillan, New York.
DHIA Sire Summary List (annual volumes). ARS, USDA. U.S. Government Printing Office, Washington, D.C.
French, M. H., I. Johansson, N. R. Joshi, and E. A. McLaughlin. 1966. *European Breeds of Cattle,* Vols. I & II. Food and Agriculture Organization of United Nations, Rome.
Perry, E. J. 1960. *The Artificial Insemination of Farm Animals,* 3rd rev. ed. Rutgers University Press, New Brunswick, N.J.
Porter, A. R., J. A. Sims, and C. F. Foreman. 1965. *Dairy Cattle in American Agriculture.* Iowa State University Press, Ames, Iowa.
Sanders, A. H. 1925. The taurine world. *National Geographic Magazine,* 48(6).
USDA, Extension Service. 1970. *Guidelines for Uniform Beef Improvement Programs* (Beef Improvement Federation Recommendation). U.S. Government Printing Office, Washington, D.C.
Review and Album (annual volumes). International Livestock Exposition, Chicago.
Breed magazines and other cattle and livestock publications.

TABLE 2.3. Summary of Cattle Breeds

Breed	Place of Origin	Date Introduced to North America	Performance Traits	Distinguishing Characteristics	Official Registration
Widely established dairy and dual-purpose					
Ayrshire	Scotland	1822	*M. wt. 1,200 1,850 B. wt. 70 80 †11,112 3.9 434	White, red, mahogany, or brown or a combination of these colored shades with white; usually red and white. Horns wide, upswept, but usually dehorned. Also polled strains.	Ayrshire Breeders' Ass. Brandon, Vt. 05733. Formed in 1875. Closed herdbook, with an open book procedure added in 1970. Identity by photos, sketches, or ear tattoos. AI accepted. About 900,000 registered to 1971.
Brown Swiss	Switzerland	1869—155 head imported	M. wt. 1,400 2,000 B. wt. 90 100 12,203 4.0 490 Gestation period about 10 days longer than other breeds.	Solid brown, from grayish or cream to very dark; little white tolerated. Horns moderately large, upswept, but usually dehorned. Also polled strains.	Brown Swiss Cattle Breeders' Ass., Beloit, Wis. 53512. Formed in 1880. Closed herdbook, with an open book added in 1969. Identity by ear tattoos. AI accepted. Over 700,000 registered to 1971.

* M. wt. = mature weight (lb.) for females and males respectively.
  B. wt. = birth weight (lb.) for females and males respectively.
† Average DHIA 305-day, 2x, mature equivalent records for both registered and grade cows of the breed calving in 1967–68, listed as lb. milk, % fat, and lb. fat, respectively.

TABLE 2.3 *(continued)*

Breed	Place of Origin	Date Introduced to North America	Performance Traits	Distinguishing Characteristics	Official Registration
Guernsey	Island of Guernsey	1831	M. wt. 1,100 1,700 B. wt. 65 75 4.6 447 9,632 Golden yellow milk fat, due to carotene content.	Golden fawn, usually with white markings; flesh (or yellowish) pigmentation preferred in muzzle and other exposed skin areas. Ideally dehorned. Also polled strains.	American Guernsey Cattle Club, Peterborough, N.H. 03458. Formed in 1877. Closed herdbook, with an open book added in 1970. Identity by photos, sketches, or ear tattoos. AI accepted. Over 3 million registered to 1971.
Holstein-Friesian	Netherlands	1621. Present cattle traceable to 1852 or later.	M. wt. 1,500 2,200 B. wt. 90 100 3.6 503 13,943	Black and white; some restrictions on amount and location of black coloring. Ideally dehorned. Also polled strains.	Holstein-Friesian Ass. of America, Brattleboro, Vt. 05301. Formed in 1885. Closed herdbook. Identity by photos or sketches. AI accepted. Over 9 million registered to 1971. Over 80% of U.S. dairy cattle are registered or grade Holsteins.
Red and White Holsteins				Recessive red mutants from purebred Holsteins.	Registry provided for by the parent Holstein Association starting in 1970. Recognized slightly earlier by Red and White Dairy Cattle Association.

TABLE 2.3 (*continued*)

Breed	Place of Origin	Date Introduced to North America	Performance Traits				Distinguishing Characteristics	Official Registration
Jersey	Island of Jersey	1850. Extensive continued importations up to recent years.	M. wt. B. wt. 8,853	1,000 50 5.0	1,500 60 443		Fawn, with or without white markings; may range from grayish or cream to almost black; dark skin & muzzle pigmentation. Refined, incurving horns, but preferably dehorned. Also polled strains.	American Jersey Cattle Club, 1521 East Broad St., Columbus, Ohio 43205. Formed in 1868. Closed herdbook. Identity by ear tattoos. AI accepted. Over 3.5 million registered to 1971.
Milking Shorthorn (Often referred to as Durhams, and registry under this name is provided for at owner's option since 1970.)	England & Scotland	1783 and earlier	M. wt. B. wt. 9,982	1,500 70 3.7	2,200 80 371		Red, white, red & white, or roan. Preferably dehorned. Also polled strains.	American Milking Shorthorn Society, 313 So. Glenstone Ave., Springfield, Mo. 65802. Formed in 1910. Functioned until 1948 as part of American Shorthorn Ass. Grading-up registry provided for. Identity by ear tattoos. AI accepted. Red & White Dairy Cattle Ass. is outgrowth of appendix registry established by American Milking Shorthorn Society for infusing outside red and white genetic material into the breed.

TABLE 2.3 (continued)

Breed	Place of Origin	Date Introduced to North America	Performance Traits		Distinguishing Characteristics	Official Registration
Widely established beef only						
Angus (or Aberdeen-Angus)	Scotland	1873. Earlier to Canada	M. wt. 1,400 2,000 B. wt. 65 70		Solid black; some tolerance for white in under-line. Polled.	American Angus Ass., 3201 Frederick Blvd., St. Joseph, Mo. 64501. Formed in 1883. Closed herdbook. Identity by ear tattoos. Tight AI restrictions lifted in 1972. Over 7 million registered to 1971.
Red Angus (Recessive red mutant from Black Angus)			Same as Angus		Solid red; Some tolerance for white in under-line. Polled.	Red Angus Ass. of America, Box 776, Denton, Tex. 76201. Formed in 1954. Identity by ear tattoos. Weaning weights required for registry. AI accepted.
Brahman (American blend of several Zebu breeds brought direct from India, or via imports from Brazil; called Cebú in South America.)	India (*Bos indicus* type)	1849	M. wt. 1,200 1,800 B. wt. 70 75 Highly adaptable to hot, humid, insect-infested areas, either pure or in crosses with *Bos taurus* types		Characteristic hump at top of shoulders. Usually dark gray color in males and lighter gray in females; any color except brindle acceptable. Large, drooping ears. Much loose skin in neck, brisket, and navel area.	American Brahman Breeders' Ass., 4815 Gulf Freeway, Houston, Tex. 77023. Formed in 1924. Closed herdbook. AI closely restricted.

TABLE 2.3 (continued)

Breed	Place of Origin	Date Introduced to North America	Performance Traits	Distinguishing Characteristics	Official Registration
Charolais (Established in U.S. mainly by grading-up. Beef breeds specified as a basis at first, but now either beef or dairy breeds acceptable.)	France	1936 from herd established in 1930 by Jean Pugibet in Mexico.	M. wt. 1,250–2,200 2,200–2,800 B. wt. 85–90 90–95 Superior size, muscling, and rate of gain. Used in U.S. mainly for crossing with British breeds.	White, light wheat, or cream. Long, tall, trim-middled, with bulging muscles. Both polled and horned strains.	American-International Charolais Ass., 1610 Old Spanish Trail, Houston, Tex. 77002. Formed in 1957. Open herdbook. Animals of recorded parents that are 31/32 Charolais are eligible for registry as purebreds. AI used under specified regulations. Charbray Division registers cattle combining Charolais and Brahman breeding.
Hereford	Hereford County, England	1817 by Henry Clay to Kentucky	M. wt. 1,450 1,800 B. wt. 70 75 Hardy, good rustlers. Cows generally poor milk producers. Often referred to as "white faces."	Red body from yellowish to dark red; white face, crest, brisket, underline, and switch. Horns curved downward and forward. White face is inherited as a dominant.	American Hereford Ass., Hereford Drive, Kansas City, Mo. 64105. Formed in 1881 under slightly different name. Closed herdbook. Identity by ear tattoos. AI closely restricted but being liberalized in 1973.

TABLE 2.3 (continued)

Breed	Place of Origin	Date Introduced to North America	Performance Traits	Distinguishing Characteristics	Official Registration
Polled Hereford	Developed from selection of polled mutants of Hereford breed by Warren Gammon, Des Moines, Iowa, starting in 1901.		Similar to Hereford. Performance at testing stations indicates some superiority over Herefords, due to leadership in emphasizing recorded performance.	Same as Hereford except for the dominant polled trait.	American Polled Hereford Ass., 4700 E. 63rd St., Kansas City, Mo. 64130. Formed in 1900 and under present name since 1947. Closed herdbook specifies Hereford or Polled Hereford registered parents. AI closely restricted but liberalization occurring.
Shorthorn (Sometimes referred to as "Scotch" Shorthorns to distinguish from Bates [milking] strains.)	England	1783 and earlier	M. wt. 1,500 2,000 B. wt. 65 75 Modest performance and carcass merit under modern evaluation criteria. Cows are better milkers than Angus and Hereford. Tendency toward patchiness of fat covering.	Red, white, red & white, or roan; roan generally preferred in the past. Rather short, outward-growing horns.	American Shorthorn Ass., 8288 Hascall St., Omaha, Neb. 68124. Formed in 1882. Closed herdbook. Identity by ear tattoos. AI closely restricted. Over 3 million registered to 1971.
Polled Shorthorn	Developed in U.S. from polled mutants of Shorthorn breed.		Same as Shorthorn	Same as Shorthorn except for dominant polled trait.	Registered by American Shorthorn Ass., identified as polled.

TABLE 2.3 (continued)

Breed	Place of Origin	Date Introduced to North America	Performance Traits	Distinguishing Characteristics	Official Registration
American-produced breeds, from Brahman × British breed crosses					
Santa Gertrudis (Brahman X Shorthorn)	Developed at King Ranch in Texas starting in 1918. Breed carries about 5/8 Shorthorn and 3/8 Braham breeding.		M. wt. 1,600 2,200 B. wt. 70 75 Long, tall, muscular, fast-gaining. Cows are generally good milkers. Heat tolerance comparable to Brahman.	Solid cherry red: little white on underline or in switch tolerated. Flat, short horns extending outward and downward. Also polled strains.	Santa Gertrudis Breeders International, Kingsville, Tex. 78363. Formed in 1951. Open herdbook. Herd or individual registry provided, with system of inspection classification and branding. AI closely restricted.
Brangus (Brahman X Angus)	Initial crossing leading to breed's development started about 1925 by USDA at Jeanerette, La. Breed carries about 5/8 Angus and 3/8 Brahman breeding.		M. wt. 1,400–1,600 1,800–2,200 B. wt. 70 75 Longer and taller than Angus. Main asset is heat tolerance from Brahman ancestry.	Solid black. Polled.	International Brangus Breeders Ass., 908 Livestock Exchange Bldg., Kansas City, Mo. 64102. Formed in 1949 under slightly different name. Open herdbook, with recording to reach specified crosses. Inspection requirements. AI permitted.
Red Brangus				Similar to Brangus except red color.	American Red Brangus Ass., 620 Colorado, Austin, Tex. 78701

Breed	Place of Origin	Date Introduced to North America	Performance Traits	Distinguishing Characteristics	Official Registration
Beefmaster (Brahman × Hereford × Shorthorn)	Developed on Ed C. Lasater Ranch, Falfurrias, Tex., starting about 1908. Ranch later located at Matheson, Colo. Recognized as a breed in 1954.		Similar to Santa Gertrudis and Brangus. Selection emphasis has been on performance rather than color and conformation.	No color requirements. Type is intermediate between Brahman and British breeds.	Beefmaster Breeders Universal, Gunter Hotel, San Antonio, Tex. 78206. Formed in 1961. Unique recording and franchise system, requiring descent from original foundation. No pedigrees or individual registry.
Breeds not widely established Red Dane (or Red Danish)	Denmark	1936 by USDA	M. wt. 1,400 2,000 B. wt. 85 90 13,172 3.9 511 Dairy or dual-purpose type, of major importance in Denmark.	Solid red. Both horned and polled strains.	American Red Danish Cattle Ass., Marlette, Mich. 48453. Formed in 1945. DHIA records required in maternal line of pedigree. Open herdbook provides grading-up procedure. AI accepted.
Dutch Belted	Netherlands (Known as Lakenvelders)	1888. Used by P. T. Barnum in 1840 circus.	M. wt. 1,200 1,750 B. wt. 75 85 Dairy type. Performance similar to original unimproved Holsteins.	Black with white belt at least 6 in. wide around body between shoulders and hips.	Dutch Belted Cattle Ass. of America, 6000 Northwest 32nd Ave, Miami, Fla. 33159
Devon	Northern Devon County, England	1623 by the Plymouth Colony	M. wt. 1,200 1,800 B. wt. 55 60 Dual-purpose to beef type. Selection is being aimed at beef performance merit.	Solid red, dark preferred; light skin color on muzzle and around eyes. Both polled and horned strains.	American Devon Cattle Club, Goldendale, Wash. 98620. Formed in 1905. Closed herdbook. Identity by ear tattoos.

TABLE 2.3 (continued)

Breed	Place of Origin	Date Introduced to North America	Performance Traits	Distinguishing Characteristics	Official Registration
Red Poll	Counties of Suffolk and Norfolk, England	1873	M. wt. 1,200–1,500 1,800–2,000 B. wt. 65 85 Dual-purpose to beef type. Milk records, gain records and carcass evaluation have been promoted by the breed organization. Future potential is likely as a beef breed.	Solid red, dark preferred. Polled.	Red Poll Cattle Club of America, Inc., 3275 Holdridge St., Lincoln, Neb. 68503. Formed in 1883. AI accepted with some restrictions.
Galloway	Scotland	1853 to Canada	M. wt. 1,400 2,000 B. wt. 65 70 Beef type	Solid black. Long, soft, curly hair. Polled.	American Galloway Breeders' Ass., 1020 Rapid St., Rapid City, S.D. 55701
Belted Galloway	Scotland		Same as Galloway	Black or dun, with white belt around midsection. Polled.	American Belted Galloway Cattle Breeders' Ass., South Fork Station, West Plains, Mo. 65776.
Highland (or Scotch Highland)	Scotland	1883	M. wt. 900 1,200 B. wt. 60 65 Beef type. Very hardy.	Long, shaggy, hair coat usually brown or dun color; other solid colors or brindle acceptable. Long, wide-spreading horns.	American Scotch Highland Breeders' Ass. Edgemont, S.D. 57735.
Dexter	Ireland	Early 1900s	M. wt. 500–800 800–900 Dual-purpose type. Dwarfism is an established part of true-type inheritance.	Black or red color. Horned.	American Dexter Cattle Ass., Decorah, Iowa 52101.

TABLE 2.3 (continued)

Breeds introduced to America mainly since 1965. (Referred to in beef industry as "exotic" or "new" breeds)

Breed	Place of Origin	Date Introduced to North America	Performance Traits		Distinguishing Characteristics	Official Registration
Simmental (called Fleckvieh in Germany and Pie Rouge in France. Also referred to in their general distribution throughout Europe as "Spotted Mountain Cattle.")	Switzerland	1967 to Canada. First pure Simmental bull from Canada to U.S. in 1970.	M. wt. 1,450–1,800 B. wt. 2,400–2,800 90 100 Dual-purpose type. 287-day av. gestation. Milk yield slightly superior to European Brown Swiss counterparts.		Light red to cream with white face and variable amount of white areas on body.	American Simmental Ass., Inc., Box 24, Bozeman, Mont. 59715. Grading-up registry system, using AI or natural service to purebred sires.
Limousin	Aquitaine region of southwestern France.	1969 to Canada	M. wt. 1,300 2,400 B. wt. 75–80 80–90		Rich red-gold shading to buckskin or tan under belly, on legs, and around muzzle. Horned.	North American Limousin Foundation, 1140 Delaware St., Denver, Colo. 80204. Formed in 1968. Weaning and post-weaning performance records required for registry. Grading-up program. AI accepted.
Maine-Anjou	Haut-Anjou and Main-et-Loire regions of France from crossing Durham and Manceau (local) cattle.	1969 to Canada	M. wt. 1,800 3,000 B. wt. 95 100		Predominantly dark red, with variable amount of white spotting. One of the largest breeds. Horned.	International Maine-Anjou Ass., Livestock Exchange Building, P.O. Box 5636, Kansas City, Mo. 64102. Formed in 1970. Grading-up program. AI accepted.

TABLE 2.3. *(continued)*

Breed	Place of Origin	Date Introduced to North America	Performance Traits	Distinguishing Characteristics	Official Registration
Murray Grey	Australia, from a cross between Angus and Shorthorn, beginning in 1905.	1969. Semen only to U.S.	M. wt. 1,300–1,500 2,000 & over B. wt. 60 70	Silver-gray with dark skin pigmentation. Predominantly polled.	American Murray Grey Ass., Inc., Route 4, Shelbyville, Ky. 40065. Grading-up system using imported semen.
Hays Converter	Canada on ranch of Harry Hays, Calgary, Alberta, by crossing Hereford, Holstein, and Brown Swiss.	1969 by American Breeders Service for AI use.	M. wt. 1,400 2,400–3,000 B. wt. 80 95 Average daily gain of 3.68 lb. in 140-day tests, and 365-day wt. of 1,135 lb. exceeded all other breeds tested in 1967–68. Canadian Government supervised R.O.P. tests. Nearest competitor was Charolais.	Most are black with white faces and underlines, dark skin pigmentation. Color has not been emphasized. Horned.	None established.
Lincoln Red	England	A few in Canada as of 1970.	Similar to Devon or Durham	Solid red. Both polled and horned strains.	Not established in U.S.
Chianina	Chianina Valley of central Italy.	1971 to Canada	M. wt. 1,800 3,000–4,000 B. wt. 110 130 Initially developed for draft purposes since ancient Roman times. Now used primarily for meat. Reputed to be world's largest cattle.	Solid white, with black skin pigmentation. Horned.	American Chianina Ass., Inc., Box 11537, Kansas City, Mo. 64138. Formed in 1971.

TABLE 2.3. *(continued)*

Breed	Place of Origin	Date Introduced to North America	Performance Traits	Distinguishing Characteristics	Official Registration
South Devon	England	1969	Beef performance being evaluated on importation of bulls to U.S. Closely comparable to Devon	Solid medium to yellowish red. Both polled and horned.	Currently being promoted by Big Beef Hybrids, P.O. Box 248, Stillwater, Minn. 55082.
Blonde d'Aquitaine	France	Semen only to U.S. in 1971 (first to be brought in under new U.S. regulations which became law in 1965)	M. wt. 1,600–1,800 2,600 & over B. wt. 110 115 Likely shares common elements with Charolais, Limousin, and Simmental breeds.	Yellow-brown, fawn, or wheat. Horned.	Blonde d'Aquitaine Society of America, P.O. Box 250, Stillwater, Minn. 55082.
Welsh Black	Wales	1969 to Canada	M. wt. 1,200–1,700 1,800–2,500 Developed along dual-purpose lines, with recent emphasis mainly on beef.	Solid black. Horned.	Canadian Welsh Black Cattle Society, Taber, Alberta, Canada. Currently serves both Canadian & U.S. interests. Grading-up program through use of AI.
Pinzgau	Pinzgau Valley in Austria	1972 to Canada	M. wt. 1,300–1,650 2,200–2,900 B. wt. 110 120 Developed for draft, meat, and milk. Of interest for beef production in North America.	Chestnut brown, with unique white marking over topline, belly, thighs, elbows, and brisket.	None established.

Additional "exotic" introductions not listed above include: Gelbvieh, a yellow German dual-purpose breed that apparently represents elements common to the Simmental and Brown Swiss breeds; Marchigiana, a large white Italian breed similar to the Chianina; and Australian Illawarra Shorthorn (A.I.S.), a dairy type Shorthorn being introduced via frozen semen into the Milking Shorthorn breed.

FIG. 2.8. (left) *Ayrshire cow*. (right) *Ayrshire bull*. (Ayrshire Breeders Assn.)

FIG. 2.9. (left) *Brown Swiss cow*. (right) *Brown Swiss bull*. (Brown Swiss Cattle Breeders Assn.)

FIG. 2.10. (left) *Guernsey cow*. (right) *Guernsey bull*. (American Guernsey Cattle Club)

FIG. *2.11.* (left) *Holstein-Friesian cow.* (right) *Holstein-Friesian bull.* (Holstein-Friesian Assn. of America)

FIG. *2.12. Red and White Holstein cow—painting of actual cow.* (Larry Moore, Suamico, Wis.)

FIG. *2.13.* (left) *Jersey cow.* (right) *Jersey bull.* (American Jersey Cattle Club)

FIG. *2.14.* (left) *Milking Shorthorn cow.* (right) *Milking Shorthorn bull.* (American Milking Shorthorn Society)

FIG. *2.15.* (left) *Angus cow; 15 progeny include 11 males with av. 205-day weaning weight of 593 lb.* (right) *Angus bull.* (Wye Plantation, Queenstown, Md.)

FIG. *2.16. Red Angus bull, cow, and calf.* (Red Angus Association of America)

FIG. *2.17.* (left) *Brahman cow.* (right) *Brahman bull.* (American Brahman Breeders Assn.)

FIG. *2.18. Charolais bull.* (American International Charolais Assn.)

FIG. *2.19. Charbray cow.* (American International Charolais Assn.)

FIG. *2.20.* (left) *Hereford cow and calf.* (right) *Hereford bull.* (American Hereford Assn.)

FIG. 2.21. *Polled Hereford bull in show condition.* (American Polled Hereford Assn.)

FIG. 2.22. *Shorthorn cow* and *Polled Shorthorn bull.* (American Shorthorn Assn.)

FIG. 2.23. (left) *Santa Gertrudis cow.* (right) *Santa Gertrudis bull.* (Santa Gertrudis Breeders International)

FIG. 2.24. *Brangus bull.* (International Brangus Breeders Assn.)

FIG. 2.25. *Red Brangus cow and calf.* (Paleface Ranch, Tex.)

FIG. *2.26. Beefmaster cow.* (Lasater Ranch, Colo.)

FIG. *2.27. Red Dane cow.*

FIG. *2.28. Dutch Belted cow.*

FIG. *2.29. Devon bull.* (Devon Cattle Assn. Inc.)

FIG. *2.30.* (left) *Red Poll cow.* (right) *Red Poll bull.* (Red Poll Cattle Club of America)

FIG. 2.31. *Galloway bull (17 mo. old).*

FIG. 2.32. *Belted Galloway bull.*

FIG. 2.33. *Highland bull.*

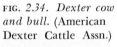

FIG. 2.34. *Dexter cow and bull.* (American Dexter Cattle Assn.)

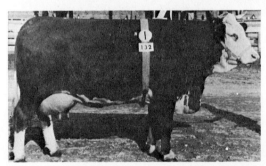

FIG. 2.35. (left) *Simmental cow (Germany).* (American Simmental Assn.) (right) *Simmental bull.* (Curtiss Breeding Service)

FIG. 2.36. *Limousin bull.*
(Curtiss Breeding Service)

FIG. 2.37. *Maine-Anjou bull.*
(International Maine-Anjou Assn.)

FIG. 2.38. (left) *Murray Grey cow.* (right) *Murray Grey bull.* (Murray Grey, U.S.A., Inc.)

FIG. 2.39. (left) *Hays Converter cow and calf.* (right)
Hays Converter bull. (American Breeders Service, Inc.)

FIG. 2.40. (left) *Lincoln Red cow.* (right) *Lincoln Red bull.* (Donald Shaver, Canada)

I IG. 2.41. *Chianina bull.* (American Chianina Assn.)

FIG. 2.42. (left) *South Devon bull.* (right) *South Devon dairy herd in England.* (Big Beef Hybrids)

FIG. *2.43.* (left) *Blonde d'Aquitaine cow and calf.* (right) *Blonde d'Aquitaine bull.* (Big Beef Hybrids)

FIG. *2.44. Pinzgau cow.*

[CHAPTER THREE] Swine [PORCINES]

SWINE are among the oldest of our domesticated animal species, dating back more than 8,000 years along with the dog, goat, and horse. The ability of swine to quickly furnish man with high-quality food nutrients—under conditions of subsistence agriculture utilizing table scraps and gleanings as well as in large-scale intensive production units utilizing abundant high-energy feeds—has made them important worldwide. Swine are produced in large numbers throughout the world, especially in temperate climates where there are large human populations.

Swine, commonly called hogs or pigs, belong to the genus *Sus* of the suborder Suiformes in the order Artiodactyla. Three species are listed within this genus, all of which may have contributed to present-day breeds: *Sus scrofa* (the wild hog of continental Europe, probably the main contributor), *Sus indicus* (the wild hog of China, Japan, and Asia), and *Sus wadituaneus* (the wild grayish-black hog of Italy). The only similar animal native to the Western Hemisphere is the peccary, which belongs to a distinctly different family.

The leading countries in swine production are: (1) China, (2) Brazil, (3) United States, (4) USSR, (5) West Germany, (6) Poland, (7) Philippines, (8) Mexico, (9) France, and (10) Denmark. The total number of swine in the world is estimated at close to 500 million.

On January 1, 1970, there were 56.7 million head of swine in the United States, with heaviest concentration in the Corn Belt where climate and plentiful high-energy feed favor their production. The leading states are: (1) Iowa, (2) Illinois, (3) Indiana, (4) Missouri, and (5) Minnesota.

An inventory of swine is less useful than for cattle and sheep, since sharp fluctuations can easily occur in response to the hog:corn price ratio. Furthermore, most pig production is from gilts which are frequently sent to slaughter after producing one or two litters.

The relative proportions of fat and lean in the pork carcass provide the basis for the traditional classification of types and breeds, namely lard-type and bacon-type. Early emphasis in the United States was on the lard-type, produced under a liberal high-energy feeding system; Denmark, England, and Canada have emphasized the leaner, bacon-type hogs produced under a more limited feeding system. The latter type features production of the famous Wiltshire sides, a highly desirable lean bacon preferred in England. A decreasing demand for lard, due to competition from vegetable fats and oils for use in cooking and bak-

THE CHANGING PROFILE OF TYPE

FIG. *3.1. Lard-type show winner, American ideal of early 20th century.*

FIG. *3.2. Bacon-type show winner, a favorite in Europe and England and used in American transition from lard-type to modern meat-type.*

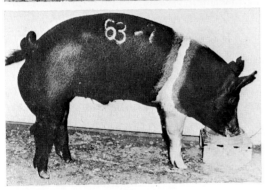

FIG. *3.3. Meat-type modern show winner (side and top-rear views show bulging muscularity and trimness of form).*

ing, changed selection emphasis in the United States from lard-type to meat-type. As often happens during periods of changing ideals, the selection pendulum initially swung sharply away from the lard-type to an upstanding type noteworthy for neither lard nor meat production, disparagingly referred to as the "meatless wonder." The meat-type ideal subsequently evolved in the industry, aided by the introduction of the live backfat probe by L. N. Hazel of Iowa State University and by increasing carcass evaluation. The modern meat-type exhibits maximum muscling and minimum fat thickness, contributing to high yields of all the preferred lean cuts from full-fed, efficient-gaining pigs, contrasted to emphasis on quality bacon alone.

PERFORMANCE EVALUATION AND IMPROVEMENT

Breed registry organizations for swine developed along the same lines as those for other species of livestock introduced into the United States. Ear notching is the standard identification system. Their promotion emphasis frequently shared with other breed organizations the shortcoming of overemphasis on fancy points of show-ring concern and slow response to commercial needs. The usual schemes of linebreeding and inbreeding were employed to fix breed characteristics. Along with the establishment of "trademark" characters such as the white belt of the Hampshire and the head shape and six white points of the Berkshire, commercial producers responded to certain desirable performance traits that became incorporated in the pure breeds. The thriftiness and prolificacy of the Duroc, for example, caused many of these red hogs to appear in midwestern feedlots.

Denmark established its first official testing station in 1907 for evaluating such traits as weight for age, efficiency of gain, and carcass quality; the East Anglian Pig Recording Scheme put a similar system into effect in England in 1927. German breeders secured data on transmitting ability in their Record of Performance program, starting about 1926. However, little evidence of recorded performance data in U.S. herds was obtained in the USDA cooperative survey of plant and animal improvement, reported in the 1937 *Yearbook of Agriculture.* Some performance recording patterned after the Danish system had been initiated by the USDA and state experiment stations about 10 years earlier, including experimentation with crossbreeding and other mating systems. In the transition from lard- to meat-type, the established bacon-type breeds such as Yorkshire, Landrace, and Tamworth were introduced and evaluated as purebreds and in crosses with existing breeds.

It soon became apparent that heterosis obtained from various crossbreeding systems yielded significant improvement in productive and reproductive traits and in carcass merit. It was during this time that the crossing of inbred lines to produce the Minnesota No. 1, Minnesota No. 2, Montana No. 1, and Beltsville No. 1 resulted in a period of popularity for these new hybrid breeds. Perhaps more importantly, this flurry of hybridization emphasis stimulated the breed registry organizations to take aggressive action in promoting performance testing to improve the merit of the purebreds. This is embodied in the Production Registry Program, involving certified litters and certified sires, which has been rather uniformly adopted with minor variations by the swine registry organizations. These programs have accelerated rapidly and place the swine industry in a leadership position among the meat animal species with respect to genetic progress. The all-breed standards for certification of performance-tested litters, effective January 1, 1970, were:

Days to 220 lb. .... 180 or less
Carcass length .... 29.5 in. or more
Backfat thickness .... 1.5 in. or less
Loin-eye area .... 4.5 sq. in. or more

Swine testing stations were established in many states through the cooperative effort of state extension services and swine producers for uniform evaluation of performance of boar pigs and carcass merit of sib barrows, followed by sales of qualifying boars. Illustrating the progress achieved by this testing and indexing system, the testing station at Ames, Iowa, started in 1956 and over the first 10 years recorded the following improvements in carcass traits of barrows slaughtered: backfat thickness decreased 0.3 in., percentage of ham and loin increased about 8, and loin-eye area increased about 1.0 sq. in. Table 3.1 lists some current swine testing station results.

TABLE 3.1. Pigs Tested in 1970 Spring Test at Iowa Swine Testing Station, Ames

	Boars			Carcass Data on Sib Barrows			
Breed	Av. daily gain (lb.)	Feed/100 lb. gain (lb.)	Backfat probe (in.)	Length (in.)	Backfat (in.)	Loin-eye area (sq. in.)	Percent ham & loin
Chester White	2.01	244	.82	29.5	1.28	4.4	42.9
Duroc	2.06	248	.78	29.5	1.20	4.7	43.9
Hampshire	2.10	246	.72	30.0	1.09	5.0	44.5
Landrace	2.08	253	.84	29.4	1.40	4.7	42.7
Poland China	2.03	248	.75	29.5	1.16	5.2	44.2
Spotted	2.08	249	.79	29.4	1.28	4.6	43.3
Yorkshire	2.12	240	.78	30.4	1.17	4.8	43.7
Average	2.07	246	.76	29.5	1.18	4.79	43.8

Accompanying the steady improvement in the merit of purebreds available for crossing, there emerged hybrid swine breeding organizations employing a hierarchial structure similar to the poultry and corn seed stock producers. These involve a nuclear breeding herd in which the crossbreeding and performance testing are directed by a professional animal breeding specialist. Breeding stock is produced in selected "multiplier herds" according to the established breeding plan, and qualifying boars thus produced are sold to commercial producers. The ingredients of the hybrid stock are not made known to the public, and selection of boars sold to commercial producers is made by the hybrid breeding organization. It is estimated that crossbreeding is involved in over 80% of the market hogs produced in the United States, either through use of boars from the hybrid breeding organizations or through the producer's own crossbreeding of purebred stock. The ratio of lean to fat in the carcasses of hogs marketed has increased markedly in recent years; productive traits such as litter size and livability, growth rate, and feed conversion efficiency have likewise responded to the improved selection and mating systems.

LIVE ANIMAL AND CARCASS COMPETITION

The show-ring has been heavily patronized by swine breeders as by other livestock exhibitors. Changes in ideals are much more quickly evidenced in swine shows than in other farm mammals, due to the shorter generation interval and higher reproductive rate compared to cattle, sheep, or horses. No single breed or cross has dominated the market and carcass class winnings at major swine shows over the past half-century as has been true with beef cattle and sheep.

Exhibition of carcasses or retail cuts of pork has been common for a longer period of time than for beef and lamb. Ham and bacon shows have been conducted since early in this century, made possible because of the relatively nonperishable nature of these cured products. As organized performance testing got into full swing and refrigeration facilities became available at shows, the combination of on-the-hoof and carcass exhibitions became popular at regularly scheduled fairs or swine specialty shows.

Once the carcass traits of the meat-type hog were clearly defined, well-qualified swine judges have been remarkably proficient in identifying meatiness in live swine. Evidences of muscling and freedom from excess fat are more apparent in the live hog than in market cattle or lambs. This has complemented performance testing in furthering the selection of superior breeding stock.

The merit of modern U.S. swine is recognized throughout the world, resulting in many sales of breeding stock to countries around the globe. It is not unusual for individuals from improved herds to gain 2 lb. per day from weaning to slaughter or breeding age, require as little as 2.5 lb. feed per pound of gain, and yield carcasses with 6 sq. in. or more loin-eye area and 40% or more ham and loin.

FUTURE PROGRESS AND PROBLEMS

Old problems remain to be solved, and new ones appear. Even though litters of 10 or more pigs farrowed are quite common, the combination of genetic and management factors has failed to significantly increase the average number of pigs saved per litter, which has stood at slightly over 7 for many years. With the attainment of very muscular hogs with thin backfat there have appeared the porcine stress syndrome (PSS) and pale, soft, exudative lean (PSE). The former is manifested by shock and usually death when pigs at about slaughter weight or larger are stressed. The latter is undesirable from the standpoint of meat quality. Arthritis and other structural unsoundnesses also appear to be increasing concerns under intensive confinement-rearing systems. Research efforts are being directed toward solution of these problems.

READING LIST

McPhee, H. C., and O. G. Hankins. 1936. Swine—Some current breeding problems. *USDA Yearbook of Agriculture, 1936.* U.S. Government Printing Office, Washington, D.C.
Breed magazines and general swine publications.

TABLE 3.2. Summary of Swine Breeds

Breed	Place of Origin	Date Introduced to U.S.	Performance Traits			Distinguishing Characteristics	Official Registration
Most important							
Berkshire	Counties of Berkshire, Wiltshire, and Gloucestershire in south central England	1823	*M. wt.	600–800	700–900	Meat-type. Black with six white points. Erect ears, tilted forward. Dished face with slightly upturned snout.	American Berkshire Ass., 601 Monroe St., Springfield, Ill. 62704. Formed in 1875. *American Berkshire Record* published in 1876 was first swine herdbook in the world. Close to 1 million head have been registered to 1971.
			Litter size: 7–10 Intermediate in performance and above average in carcass traits.				
Chester White	U.S., in Chester, Delaware, Lancaster, and Philadelphia counties in Pennsylvania. Other strains included O.I.C. (Ohio Improved Chester) developed in Ohio. From crosses of several breeds, including Yorkshire		M. wt.	650–850	750–950	Meat-type. White. Medium to large drooping ears. Head moderately large with profile nearly straight.	Chester White Swine Record Ass., Box 228, Rochester, Ind. 46975. Formed in 1884. Well over 1 million have been registered to 1971.
			Litter size: 9–11 Intermediate in performance and carcass traits.				
Duroc	Eastern U.S. about 1870 by combining Jersey Red from New Jersey and Duroc from New York. Formerly called Duroc-Jersey. Name Duroc comes from name of Thoroughbred stallion owned by early breeder of these red hogs, Harry Kelsey of New York.		M. wt.	700–850	800–1,000	Meat-type. Solid red from light golden to very dark. Ears medium to small, drooping downward and forward off face. Medium length, dished face.	United Duroc Swine Registry, Duroc Building, Peoria, Ill. 61602. Formed under slightly different name in 1934 from mergers of preceding organizations dating back to 1882. *Duroc Bulletin* (now *Duroc News*), started in 1904, was first magazine in the world devoted to single swine breed. About 4½ million registered to 1971.
			Litter size: 8.5–10.5 Hardy, good mothering ability and growth traits. Intermediate and rapidly improving in carcass traits.				

TABLE 3.2. *(continued)*

Breed	Place of Origin	Date Introduced to U.S.	Performance Traits	Distinguishing Characteristics	Official Registration
Hampshire	Kentucky about 1890, probably from white-belted hogs brought to America around 1830 from Hampshire County, England. Initially known as Thin Rind and other local designations before agreeing on Hampshire in 1904.		M. wt. 550–700 650–850 Litter size: 8.5–10.5 Good milking and mothering ability. Intermediate performance and trim, meaty carcasses.	Meat-type of leaner background than old line U.S. breeds. Black with white belt extending from one front foot and leg to the other. Small, erect ears; moderately long, trim head with well-dished face.	Hampshire Swine Registry, 111 Main St., Peoria, Ill. 61602. Parent organization formed in 1893. About 2 million registered to 1971.
Landrace	Denmark about 1895 from crossing Large White from England with native stock. Small amount of Poland China breeding introduced in development of American Landrace.	1934 by USDA for experimental testing	M. wt. 550–750 700–900 Litter size: 9–11 Prolific, good mothers, but pigs lack hardiness of older breeds in America. Danish imports were bacon-type, reflected now in long, lean carcasses.	Meat-type, from bacon-type original stock. White. Long body with low, flat arch of back. Large, drooping ears that lie forward more closely along sides of face than Chester White; head, long & narrow with straight profile.	American Landrace Association, Culver, Ind. 46511. Formed in 1950. Over 150,000 registered to 1971.
Poland China	Butler and Warren counties in southwestern Ohio, about 1850, from stock of Russian, Chinese, Irish, and mixed background with later addition of Berkshire.		M. wt. 650–950 850–1,000 Litter size: 7.5–9.5. Has undergone several type changes during its history, including short, round "hot blood" type of late 1800s. Current type excels most breeds in meatiness and is about average in other performance traits.	Heavily muscled, meat-type. Black with 6 white points including face blaze. Ears medium in size, drooping, but carried well away from the face. Dished face and wide-set eyes.	Poland China Record Ass, 501 E. Losey St., Galesburg, Ill. 61404. Formed in 1905 following several parent organizations dating back to 1878. About 3 million registered to 1971.

TABLE 3.2. *(continued)*

Breed	Place of Origin	Date Introduced to U.S.	Performance Traits	Distinguishing Characteristics	Official Registration
Spotted (or Spot)	Putnam and Hendricks counties in Indiana, about 1914, from native stock of Poland China background, Gloucester Old Spots, and the then-recognized Spotted Poland China. Formerly called Spotted Poland China.		M. wt. 650–850 750–950 Litter size: 8.5–10.5 Above average in most performance and carcass traits. Very similar to Poland China.	Meat-type. Black and white with some distinct spots; either color must be within limits of 20 and 80% on body; legs usually white. Small ears, drooping downward and forward, but held away from face. Straight or slightly dished face.	National Spotted Swine Record, Inc., Bainbridge, Ind. 46105. Formed in 1914. Over 1 million registered to 1971.
Yorkshire	Yorkshire County and surrounding area in northern England. Three types classed in England as Small Whites, Middle Whites, and Large Whites.	1860 (Small Yorkshires). Later importations of larger types from England and Canada established the basis for present Yorkshires in U.S.	M. wt. 600–800 750–1,000 Litter size: 9.5–11.5 Good mothering ability, and unexcelled for size of litters. Not as hardy as some other breeds under average U.S. conditions. Bacon-type background results in leanness of carcass and high feed efficiency.	Meat-type, developed from bacon-type stock. Conformation resembles bacon-type. Solid white. Erect ears. Moderate to strongly dished face.	American Yorkshire Club, 1001 South Street, Lafayette, Ind. 47902. Formed in 1893. About 600,000 registered to 1971.

TABLE 3.2. (continued)

Breed	Place of Origin	Date Introduced to U.S.	Performance Traits	Distinguishing Characteristics	Official Registration
Less important Hereford	Missouri, on farm of R. V. Webber about 1902. Further development by Iowa and Nebraska breeders in early 1920s. From crossing Duroc, Chester White, and Poland China.		M. wt. 550–750 600–800 Litter size: 7.5–10.0 Little performance information available. Probably less advanced than current breeds.	Red with white markings similar to Hereford cattle, from which it got its name. Ear carriage similar to parent stock. Meat-type.	National Hereford Hog Record Ass., R.R. 3, Shelbyville, Ill. 62565. Formed in 1934. Over 100,000 have been registered, with current annual registration of 500–600.
Lacombe	Canada Dept. of Agriculture at Lacombe Exp. Farm, Lacombe, Alta., starting in 1947. From crossing Chester White, Landrace, and Berkshire. Herdbook closed in 1952.	About 1962	M. wt. 500–800 600–900 Litter size: 10 Little performance information available. Probably intermediate to parent breeds.	Meat-type, resembling Landrace. White. Drooping ears, about intermediate to Landrace and Chester White in size and carriage.	American Lacombe Swine Ass., Grand Mound, Iowa 52751. Littermate performance data required for registration of boars.
Large Black	Somerset, Devonshire, and Cornwall counties in southwest England about 1850.	1910	M. wt. 500–800 600–900 Litter size: 9.5–11.5 Not officially performance tested in U.S. testing stations. Prolific and popular breed in England.	Solid black. Bacon-type. Very large drooping ears carried close to sides of face.	National Large Black Swine Breeders Ass., Midland, N.C. 28107. Formed in 1959. Association recognizes "Blue Spotted Hybrids," which result from mating Large Black with either Landrace or Yorkshire.
Tamworth	Stafford County, England, starting in 1812, from Irish Grazer breed imported from Ireland. In early years referred to as "Grizzly" or "Mahogany."	About 1881	M. wt. 550–750 700–900 Litter size: 8.5–10.5 Good mothering ability. Performance and carcass traits comparable to other breeds of bacon-type background.	Solid golden to cherry red. Meat-type, developed from bacon-type. Erect ears. Long narrow head. Lean, thin body.	Tamworth Swine Ass., R.R. 2, Cedarville, Ohio 45314. Formed in 1897 under slightly different name. About 200,000 registered to 1971.

TABLE 3.2. (continued)

Breed	Place of Origin	Date Introduced to U.S.	Performance Traits	Distinguishing Characteristics	Official Registration
Wessex Saddleback	England, in Dorsetshire County area, before 1800.	1825	M. wt. 500–700 600–800 Litter size: 9–10 Limited performance data available in U.S. indicate below-average performance and carcass traits, except above-average prolificacy. Used in crosses to develop Hampshire breed.	Black body with white belt like Hampshire. Large, drooping ears distinguish it from Hampshire. Meat-type developed from bacontype, which it resembles in body conformation.	National Wessex Saddleback Swine Ass., Box 2145, Des Moines, Iowa 50310. Formed in 1955.
"Hybrid" breeds developed by crossing inbred pure lines.					
Beltsville No. 1 (Danish Landrace & Poland China crosses)	USDA	Started 1934	Comparable to many recent hybrids.	Black with white spots. Large drooping ears like Landrace.	Inbred Livestock Registry Ass., Box 312, Augusta, Ill. 62311
Beltsville No. 2 (Danish Landrace, Yorkshire, Duroc & Hampshire crosses)	USDA	Started 1940	Comparable to many recent hybrids.	Red with whitish underline; some black spotting. Erect ears.	Inbred Livestock Registry Ass., Box 312, Augusta, Ill. 62311
Maryland No. 1 (Berkshire & Landrace crosses)	Blakeford Farms, Queenstown, Md. (better known for Angus and Guernsey cattle), Md. Agr. Exp. Station and USDA.	Started about 1940	Probably comparable to many recent hybrids.	Black body with white spots and markings. Small, usually erect ears. Resembles Berkshire.	Inbred Livestock Registry Ass., Box 312, Augusta, Ill. 62311

TABLE 3.2. *(continued)*

Breed	Place of Origin	Date Introduced to U.S.	Performance Traits	Distinguishing Characteristics	Official Registration
Minnesota No. 1 (Tamworth & Danish Landrace crosses)	Minnesota Agr. Exp. Station and USDA.	Started about 1935	Probably comparable to many recent hybrids.	Solid red; may have black spotting. Long body. Large drooping ears.	Inbred Livestock Registry Ass., Box 312, Augusta, Ill. 62311
Minnesota No. 2 (Yorkshire & Poland China crosses)	Minnesota Agr. Exp. Station and USDA.	Early 1940s	Probably comparable to many recent hybrids.	Black with white markings. Usually erect ears.	Inbred Livestock Registry Ass., Box 312, Augusta, Ill. 62311
Minnesota No. 3 (From crossing 8 breeds)	Minnesota Agr. Exp. Station and USDA.	Early 1950s	Probably comparable to many recent hybrids. Selected for fertility and production.	Variable colors. Usually drooping ears.	Inbred Livestock Registry Ass., Box 312, Augusta, Ill. 62311
Montana No. 1 (Hampshire & Danish Landrace crosses)	Montana Agr. Exp. Station and USDA.	Started 1936	Probably comparable to many recent hybrids.	Solid black. Large, drooping to partially erect ears.	Inbred Livestock Registry Ass., Box 312, Augusta, Ill. 62311
Palouse (Landrace & Chester White crosses)	Washington Agr. Exp. Station.	Started 1945	Probably comparable to many recent hybrids.	Solid white. Ear carriage variable.	Inbred Livestock Registry Ass., Box 312, Augusta, Ill. 62311

 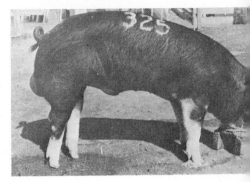

FIG. 3.4. (left) *Berkshire gilt.* (right) *Berkshire young boar.* (American Berkshire Assn.)

FIG. 3.5. (left) *Chester White gilt.* (right) *Chester White young boar.* (Chester White Swine Record Assn.)

FIG. 3.6. (left) *Duroc gilt.* (right) *Duroc young boar.* (United Duroc Swine Registry)

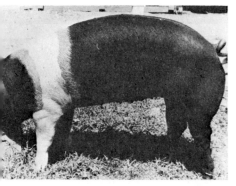 

FIG. 3.7. (left) *Hampshire gilt.* (right) *Hampshire young boar.* (Hampshire Swine Registry)

FIG. 3.8. (left) *Landrace gilt.* (right) *Landrace young boar.* (American Landrace Assn.)

FIG. 3.9. (left) *Poland China gilt.* (right) *Poland China young boar.* (Poland China Record Assn.)

FIG. *3.10.* (left) *Spotted gilt.* (right) *Spotted young boar.*
(National Spotted Swine Record, Inc.)

FIG. *3.11.* (left) *Yorkshire gilt.* (right) *Yorkshire young
boar.* (American Yorkshire Club)

FIG. *3.12.*   (left) *Hereford sow.* (right) *Hereford mature
boar.* (USDA, Bureau of Animal Industry)

FIG. *3.16.* (left) *Wessex Saddleback sow.* (right) *W
Saddleback boar.* (National Wessex Saddleback S
Assn.)

FIG. 3.13. (left) *Lacombe gilt*. (right) *Lacombe young boar.*

FIG. 3.14. *Large Black gilt.*

FIG. 3.15. (left) *Tamworth sow*. (right) *Tamworth mature boar.* (Tamworth Swine Assn.)

FIG. 3.17. *Hybrid breeds registered by Inbred Livestock Registry Assn.*

Beltsville No. 1 gilt. (USDA, Bureau of Animal Industry)

Beltsville No. 2 boar. (USDA, Bureau of Animal Industry)

Maryland No. 1 gilt.

Minnesota No. 1 gilt. (Univ. of Minn.)

Minnesota No. 2 sow and litter. (Connor Prairie Farms)

Minnesota No. 3 boar. (Univ. of Minn.)

Montana No. 1 gilt.

Palouse gilt (Wash. State Univ.)

[CHAPTER FOUR] Sheep [OVINES]

SHEEP are raised around the world on all continents and in all climates. They are raised chiefly for meat and wool, but some are milked. Cheese made from sheep's milk is very popular in many countries, including the original Roquefort cheese produced in southern France. It is known that Hebrews were eating lamb during their stay in Egypt. Sheep have been prized since very early times for their fleeces, which provided one of the few good fibers for processing into clothing. Wool remains the standard of excellence for many uses such as quality carpeting and men's suits, although synthetic fibers have increasingly challenged the traditional natural fibers. Technology has enhanced the inherent characteristics of wool to better meet modern textile demands.

Domesticated sheep, *Ovis aries,* belong to the suborder Ruminantia (true ruminants) of the order Artiodactyla (even-toed). They may also be grouped in the Bovidae family of this suborder.

There are slightly fewer than 1 billion head of sheep in the world, according to recent USDA estimates. World numbers have shown a steady increase, whereas numbers have decreased in the United States. The leading countries in sheep numbers are (1) Australia, (2) USSR, (3) China, (4) New Zealand, (5) Argentina, (6) India, (7) Turkey, (8) Ethiopia, (9) United States, and (10) Uruguay.

On January 1, 1970, there were 20.4 million head of sheep in the United States, compared to the peak number of 56 million in 1942. The leading states in sheep numbers are: (1) Texas, (2) Wyoming, (3) California, (4) Colorado, (5) South Dakota, (6) Montana, (7) Utah, (8) Iowa, (9) New Mexico, and (10) Idaho. Sheep are raised in all 50 states.

TYPES AND USES

In order of current importance, three major types of sheep are produced in the United States: (1) meat (mutton sheep), (2) wool, and (3) fur. The fur type is typified by the Karakul breed, which is actually a specialized wool breed from which pelts of lambs killed at birth or shortly thereafter are known as Persian lamb.

Breeds kept for meat and/or wool may be classified into three wool types: (1) medium-wool, producing fleeces that are medium in length, fineness, and weight; (2) long-wool, with long fibers of 5–12 in. and less crimp; and (3) fine-wool, typified by sheep of Merino or Rambouillet background producing fine, small fibers with 16–22 crimps per inch. Fine-wool fleeces have the highest shrink during scouring due to a higher content of yolk and natural grease. In wool terminology, the fleece as sheared is known as "grease wool"; after going through the scouring process to remove yolk, grease, and foreign matter it is known as "clean wool." Large amounts of lanolin obtained in the scouring process are used in the cosmetic industry. The only breeds raised in large numbers primarily for wool production in the United States are the fine-wool breeds.

Wool is graded according to fineness and length of fiber (staple length). The American system, usually referred to as the "blood system," evolved from a fine-wool base of Merino breeding and roughly indicates the percentage of Merino breeding (or "blood") in the sheep that produced the wool. These grades according to fineness are *fine, half blood, three-eighths blood, quarter blood, low quarter blood, common,* and *braid.* The British "spinning count" system is more precise and is based on the number of hanks (one hank = 560 yd.) of yarn that can be spun from one pound of clean wool. The grades are further subdivided into classes based on staple length. The three main classes are *staple, French combing,* and *clothing* wool.

A comparison of the two grading systems is provided by the USDA specifications for grade and length of fiber for the most common grades produced, together with approximate shrinkage expected for each grade.

Blood Grade	Spinning Count	Staple Length Classes (in.)			Expected Shrink (%)
		Staple	French combing	Clothing	
Fine	80's 70's 64's	> 2.75	1.25–2.75	< 1.25	56–61
½	62's 60's	> 3.00	1.50–3.00	< 1.50	53–59
⅜	58's 56's	> 3.25	2.25–3.25	< 2.25	47–53
¼	54's 50's	> 3.50	....	< 3.50	44–50
Low ¼	48's 46's	> 4.00	....	< 4.00	42–48

Common and *braid* represent a very small fraction of the wool produced and run 44's and less on spinning count.

INTRODUCTION AND DEVELOPMENT

The tri-state area of West Virginia, Ohio, and Pennsylvania in the upper Ohio River Valley was an early site of sheep introduction and development in the United States, with emphasis on a larger, modernized type of fine-wool sheep developed from imported Spanish Merino stock. Vermont was another early center of Merino emphasis. Merinos were first imported in 1793, and several other importations followed soon after 1800. These small, angular, wrinkled, dense-fleeced sheep had been developed in Spain for the production of high-quality wool on the estates of wealthy agriculturists, often in flocks of 30,000–40,000 head. They were hardy and easy to handle due to strong flocking tendencies. Flocks often traveled 200–400 miles between their summer grazing ranges in the mountains of the north and their lowland wintering pastures in the south.

Selection by American breeders soon split the Merino into Types A, B, and C—depending on body size and meatiness, fleece characteristics, and degree of skin wrinkling—with Type C departing furthest from the original type and ultimately evolving into the breed known as the Delaine-Merino (more recently just Delaine). Types A and B are no longer important in the United States. The Delaine closely resembles the Rambouillet, a large-framed, fine-wool breed developed in France from Spanish Merino stock. The Von Hohmeyer flock in Prussia, developed from French stock, was most influential as a source of imported Rambouillet stock in the United States. The Rambouillet is predominant in the background of western range flocks developed for fine-wool production. Major expansions in wool production have occurred in the United States coincident with wars—the Civil War, World War I, and World War II.

British mutton or "downs" breeds followed the fine-wool breeds into the United States during the era of wide-scale introductions of all livestock breeds in the 1800s. The Southdown, a medium-wool mutton breed of long-standing identity and prestige in England, became popular throughout the East and Midwest. This breed (like its counterpart in beef cattle, the Angus) exhibits the blocky, compact, early-fattening characteristics considered to be the zenith of meat animal excellence during the first half of the 20th century. Other British mutton breeds such as Dorset, Hampshire, Shropshire, Cheviot, Suffolk, Cotswold, Lincoln, and Romney were also among the early introductions. Using the imported breeds as a basis, breeders and experiment stations in America and other major sheep-producing countries were active during this expansion period in developing new breeds combining the desired traits of the imports. The American-produced Columbia, Montadale, Panama, and Targhee and the New Zealand-produced Corriedale are examples of well-established breeds developed from such crossings.

As the sheep industry became established nationwide, emphasis on farm flock production of milk-fed lambs from prolific, mutton-type, medium-wool breeds occurred from the Corn Belt eastward, while large-scale production of fine to medium wool from large range-managed flocks was concentrated in the West and Southwest, largely from sheep of Rambouillet or Delaine background. In the latter instance late spring lambing is practiced, and single lambs are preferred over the late winter or early spring multiple births sought in the more intensively managed farm flocks. An overlap of interests exists in the Midwest, where farm ewe flocks produce early-season, milk-fed, mutton-type lambs but feedlots finish large numbers of lambs from western range flocks for late marketing.

The market centers for products of the sheep industry are principally on the East and West coasts, with city dwellers in those areas consuming most of the lamb and most textile mills being located along the East Coast. Traditionally, grease wool was shipped East to be processed; more recently, in keeping with the trend in other livestock products, wool and mohair scouring plants have been established in the West within the major production area. Large wool-scouring plants in the Edwards Plateau region of Texas are typical examples.

The saga of the sheep industry is one of the most fascinating stories in American history—dealing with human conflict and emotion, real or imagined incompatability between animal species, and the business side of sheep production. Conflicts of the western range between the cattle barons and the sheep raisers are portrayed in motion pictures, television, and western novels. The Basque shepherds from Spain's Pyrenees Mountains—with their unique and ill-translatable language—immigrated to Utah, Idaho, and other Rocky Mountain areas to hire out to owners of large sheep spreads. Their lonely existence as devoted caretakers of these flocks is closely interwoven with the progress of the industry. In the more heavily populated areas to the east, the large losses attributable to roving dogs caused constant conflict between sheep owners and hunters or village dog owners. The dog has been both friend and foe to the sheep industry. The faithful sheep dog, as typified by well-trained Border Collies, has been of inestimable value in the protection and handling of sheep. During periods of major expansion large numbers of sheep have been handled by nonowners under a share arrangement with the owners. In the Southwest this is often referred to as the "partido" system, derived from the Spanish verb meaning divided. In the East a common arrangement was to put sheep out for keeping on "halves," in which the recipient received half the return for wool and lambs from the owner.

Somewhat out of the mainstream of the industry but worthy of mention are the Navajo sheep maintained in large numbers by the Navajo Indians in the Southwest. These long-wool sheep provide wool which is usually hand-spun and woven by the Navajos into colorful blankets and other fabrics. The fat-rumped or fat-tailed sheep of Asia are naturally short-tailed and store large reserves of fat in the region of the rump and tail. This type of sheep has not become important in the Western Hemisphere, although represented by the Notail and Tunis breeds described herein.

IDENTIFICATION AND RECORDING

Each breed of sheep has a registry association, and in a few cases more than one. Individual identification of sheep for registry purposes is usually by means of ear tattoos or numbered metal ear tags. For flock management of both purebred and commercial flocks, paint branding with a paint that scours easily from the sheared fleece is commonly employed on an individual or flock basis.

Like many other breeds of purebred livestock, some of the sheep breed registry organizations in the United States were formed before such registries were established in their countries of origin.

IMPROVEMENT EFFORTS

From the standpoint of meat production, the desired traits in sheep are not unlike those for other meat animals. Muscular conformation, growthiness, reproductive rate, mothering ability, and carcass yield and quality are economically important traits. Historically, the British Isles have been the cradle of breed development for most meat animals introduced into the Western Hemisphere, and mutton-type sheep breeds are no exception. Robert Bakewell and other early English livestock breeders based their selection and breeding efforts mainly on utilitarian values, but once these evolved to identified breeds there ensued an era of emphasis on fancy points of breed type which had little relationship to economic merit. The excessively wooled face of the Shropshire, the insistence on small wool-covered ears in the Southdown, and the various face color and head shape requirements of breeds are examples of breed identity "trademarks" which have frequently been overemphasized to the detriment of productive traits. Notwithstanding these seeming errors of judgment as viewed in retrospect, progress was achieved through the introduction of improved breeds and the era of purebred emphasis which followed. Agricultural colleges and extension services actively promoted purebred ram sales, cooperative marketing through lamb and wool pools, and other such programs. These efforts helped to upgrade the quality of flocks as well as the management and handling practices necessary to merchandise a standardized, quality product. Improvement obtainable through these means gradually reached a plateau, consumer preferences changed, and research and practical observations brought the realization that new directions of emphasis were needed. Research studies proved that open-faced ewes raised heavier lambs than those with excessive face wool; larger, more open framed ewes brought lambs to market weight more easily and efficiently than the blocky, scanty-milking Southdowns; and the long-standing recommended practice of docking and castrating market lambs is questionable.

Individual and progeny evaluation based on weaning weight has become an important extension effort. Ram testing stations and on-the-farm performance testing, as well as carcass contests and evaluation, have increasingly entered the picture during the last decade. The American Hampshire Sheep Association, Stuart, Iowa, and others have actively endorsed performance testing for improving the efficiency and quality of meat production.

The objectives in market lamb production in farm flocks are prolific, good-milking ewes that will average a 150% lamb crop or better and lambs that will efficiently reach a market weight of about 100 lb. with desirable carcass quality.

Crossbreeding or grading up has long been a part of commercial sheep production. The use of crossbreeding has accelerated in recent years. The Dorset breed—well known for mothering ability, milk production, and the unique reproductive capacity to deviate from the usual seasonal breeding pattern of other breeds—often found less favor because of its heavy horns. The appearance of polled mutants in this breed at North Carolina State University and their expansion principally in Oklahoma has renewed the interest in utilizing this breed for its desirable traits. Likewise, the industry is looking at genetic resources not previously introduced into the Western Hemisphere, such as the Finnish breed which produces litters of 3–6 lambs in contrast to the usual singles or twins of presently used breeds. In general, the trend in production of meat-type sheep is toward larger, more open faced, heavier milking, more prolific, faster growing types that lend themselves to a more intensive and efficient system of meat production.

In fine-wool production, the use of a "squeeze machine" to evaluate the density and shrink of fleeces has aided in the selection of breeding stock to maximize the yield of high-quality clean wool.

The importance of uniform fiber diameter, character, and color throughout the fleece is emphasized in wool produced by all sheep. It is noteworthy that fine-wool production emphasizes white-faced sheep, which are generally free of black fibers sometimes found in the fleeces of black-faced mutton types.

A recent effort to develop a standard program for nationwide improvement of both meat and wool production, geared to both farm flock and range production, is detailed in "National Recommendations for Uniform Selection Programs in Sheep," a report developed by the Extension Sheep Program Committee of the American Society of Animal Science.

The properly managed sheep production enterprise continues to be profitable, especially the farm flock as a supplement to other income-producing units on the farm. However, low demand in the United States for lamb and mutton compared to other meats and occupational pursuits minimizing the need for heavy woolen clothing are likely to keep sheep production in the United States in a minority position among animal enterprises.

TABLE 4.1. Summary of Sheep Breeds

Breed	Place of Origin	Date Introduced to U.S.	Performance Traits	Distinguishing Characteristics	Official Registration
Meat type (Medium-wool) Cheviot	Cheviot Hills in England-Scotland border area.	1838	* M. wt. 140 173 † F. wt. 5–7 7–10 ‡ Grade: ⅜–¼ blood	Blocky, short-legged. White face and legs, free of wool, with black nostrils and hoofs. Very stylish, alert head and ear carriage, with ears erect and high-set. Hornless.	American Cheviot Sheep Society, Inc., Box 5051, Lafayette Hill, Pa. 19444. (Formed in 1900 from two existing organizations.)
Columbia	USDA Sheep Exp. Sta., Dubois, Idaho, in 1918, from crossing Lincoln, Cotswold, Romney, and Rambouillet.		M. wt. 150–200 225–300 F. wt. 11–14 20 Grade: ⅜–¼ blood	Upstanding, White, moderately open face and legs. Ears carried outward from head, partly wool-covered. Hornless, with occasional scurs.	Columbia Sheep Breeders' Ass. of America, P.O. Box 272, Upper Sandusky, Ohio 43351. Formed in 1941.
Corriedale	New Zealand from crossing Lincoln, Leicester, and Merino in early 1900s.	1914	M. wt. 125–185 185–250 F. wt. 10–12 Up to 20 Grade: Mostly ⅜ blood	Moderately tall and long, White, moderately open face. Wool-covered poll and legs. Ears small and free of wool. Hornless, except for occasional males.	American Corriedale Ass., Inc., P.O. Box 29C, Seneca, Ill. 61360
Dorset	County of Dorset, England, in early 1800s. Formerly called Dorset Horn. Polled mutants appeared in 1940s in North Carolina.	1887	M. wt. 125–175 175–225 F. wt. 7–8 10 Grade: ⅜–¼ blood	White, wool-free face. Ears small and refined. Little wool on legs and underline. Horns in both sexes, heavy and spiraled in males. Also a polled strain. Breeding not seasonally restricted. Prolific.	Continental Dorset Club, Box 1206, Carbondale, Ill. 62901. Formed in 1898. Scoring against breed scale of points required for registry.

* M. wt. = mature weight (lb.) for females and males respectively.
† F. wt. = fleece weight (lb.) for females and males respectively.

TABLE 4.1. (continued)

Breed	Place of Origin	Date Introduced to U.S.	Performance Traits	Distinguishing Characteristics	Official Registration
Hampshire	County of Hampshire in southern England, 1857.	Before 1840, but established with imports after 1880.	M. wt. 150-200 225-300 F. wt. 7-8 10 Grade: ⅜-¼ blood	Black or dark brown face, usually wool-free. Long, outward-carried, black ears. Large head with slightly Roman nose. Hornless.	American Hampshire Sheep Ass., Stuart, Iowa 50250. Formed in 1889, under present name since 1906.
Montadale	U.S. from crossing Cheviot and Columbia, starting in 1932 by Mattingly Sheep Co., St. Louis, Mo.		M. wt. 150-174 250 F. wt. 11-14 20 Grade: ½-⅜ blood	Wool-free white face and ears with some black spots. Black nose and black hoofs preferred. Ears high-set and carried upward and forward. Hornless, except for occasional scurs.	Montadale Sheep Breeders' Ass., Inc., 61 Angelica St., St. Louis, Mo. 63147. Formed in 1945.
Notail	S. Dak. Agr. Exp. Sta. by crossing several existing breeds with fat-rumped Siberian sheep, starting in 1915.	Siberians brought in 1913.	M. wt. 130-160 180-220 F. wt. 8 10 Grade: Mostly ⅜ blood	Tail absent or nearly so. White, wool-free face and ears; ears extend outward, downward, backward. Little or no wool on lower leg. Hornless.	Records being kept at S. Dak. State Univ., Agr. Exp. Sta., Brookings, S. Dak. 57006, pending release of breeding stock to public.
Oxford	Oxford County in southern England by crossing Cotswold and Hampshire.	1846	M. wt. 175-225 250-325 F. wt. 9-13 15 Grade: Mostly ¼ blood, staple	Large, massive body. Brown to light gray or black on face and legs. Some face wool, and legs wool-covered, Large, wide, high-held head, with fairly large wool-covered ears carried outward and forward. Hornless.	American Oxford Down Ass., Eaton Rapids, Mich. 48827. Formed in 1882 under slightly different name.

TABLE 4.1. *(continued)*

Breed	Place of Origin	Date Introduced to U.S.	Performance Traits				Distinguishing Characteristics	Official Registration
Panama	U.S., from crossing Rambouillet and Lincoln on James Laidlaw ranch in Idaho, starting 1912.		M. wt. F. wt. Grade:	140 11–12 ⅜–¼ blood	225 15		White face and legs. No face wool, and lower leg nearly wool-free. Medium small, wool-covered ears, carried outward. Hornless.	American Panama Registry Ass, Rupert, Idaho 83350. Flock rather than individual registry.
Shropshire	Shropshire and Stafford counties in England.	1855	M. wt. F. wt. Grade:	130–165 8 ⅜–¼ blood	180–220 10		Low-set, blocky. Face, legs, and ears brown to almost black. Ears small, wool-covered, carried outward and forward. Legs well covered with wool. Prone to "wool blindness" due to excessive face wool, which is being reduced by selection. Hornless.	American Shropshire Registry Ass, Inc., P.O. Box 1970, Monticello, Ill. 61856. Sponsors a Performance Registry Program.
Southdown	Southeastern England, deriving its name from chalk hills called South Downs in that area. One of oldest livestock breeds with development started in 1788. Bred in 19th century by royal families of England, Russia, and France.	1803	M. wt. F. wt. Grade:	120–140 6 ½–⅜ blood	160–190 8		Small, low-set, compact, "model" meat type in first half of 20th century. Face, legs, and ears gray to mouse brown in small areas not covered by wool. Small ears woolcovered, face increasingly wool-free. Head small; face short and broad; alert head carriage. Hornless.	American Southdown Breeders' Ass, 212 South Allen St., State College, Pa. 16801.

TABLE 4.1. *(continued)*

Breed	Place of Origin	Date Introduced to U.S.	Performance Traits	Distinguishing Characteristics	Official Registration
Suffolk	Counties of Norfolk, Suffolk, Essex, and Kent in England, by crossing Old Norfolk with Southdown	1888	M. wt. 175–225 250–350 F. wt. 6–9 10 Grade: Usually ⅜ blood	Large, muscular. Face, head, ears, and legs jet black and wool-free. Head rather long and narrow, with Roman nose. Large ears extend outward with upward turn at end. Hornless. One of most popular modern sheep breeds.	National Suffolk Sheep Ass, P.O. Box 324, Columbia, Mo. 65201. American Suffolk Sheep Society, Moscow, Idaho 83843
Targhee	USDA Sheep Experiment Station, Dubois, Idaho, from crossing Rambouillet, Corriedale, and Lincoln, starting in 1926.		M. wt. 140 200–225 F. wt. 11–13 15 Grade: Low fine to ⅜ blood, averaging about ½ blood, staple.	Chiefly white in exposed areas of face, ears, and legs with black or brown markings allowed. Head and legs well wool-covered, face nearly wool-free. Hornless	United States Targhee Sheep Ass, P.O. Box 2513, Billings, Mont. 59103
Tunis	Northern Africa, ancient origin.	1799	M. wt. 130–140 150–175 F. wt. 8 10 Grade: ⅜–¼ blood	Fat-tailed type. White, tan, mottled brown, and red are common color markings on exposed areas. Some dark gray to reddish fibers in wool. Lambs tan or red at birth, later becoming white. Wool-free face, ears, and lower legs. Head small, long, narrow; ears very long, broad, pendulous. Nonseasonal breeders. Hornless.	National Tunis Sheep Registry, Inc., Bath, N.Y. 14810

TABLE 4.1. (continued)

Breed	Place of Origin	Date Introduced to U.S.	Performance Traits	Distinguishing Characteristics	Official Registration
Meat type (Long-wool)					
Cotswold	Cotswold Hills of Gloucester County, England	1832	M. wt. 200–225 / 300; F. wt. 12 / 20. Grade: Common to mostly braid, 8–10 in. length.	Large, upstanding. White face, ears, and legs, with some grayish or dark spots acceptable. Ears small, wool-covered. Lower legs usually wool-free. Long, curly ringlets of wool hang from poll and body. Hornless, but frequently scurs in males.	American Cotswold Record Ass., Sigel, Ill. 62462
Lincoln	Lincoln County, England, as early as 1749.	Late 1700s	M. wt. 225–250 / 300–325; F. wt. 12–14 / 20–25. Grade: Mostly braid, 10–15 in. length.	Largest breed. White wool-free face, with long spiral forelock wool hanging over face and eyes. Hornless.	National Lincoln Sheep Breeders' Ass., West Milton, Ohio 45383
Romney	Romney Marsh area of Kent County, England. Formerly called Romney Marsh.	1904	M. wt. 200 / 250–300; F. wt. 10–12 / 15. Grade: ¼–low ¼ blood, staple; 7–10 in. length.	White face, ears, and legs. Face wool-free except jaws. Long, curly locks of dense, compact wool. High head carriage; pointed ears set well back on sides of head.	American Romney Breeders' Ass., Corvallis, Ore. 97330

TABLE 4.1. *(continued)*

Breed	Place of Origin	Date Introduced to U.S.	Performance Traits	Distinguishing Characteristics	Official Registration
Fine-wool Debouillet	U.S. on New Mexico ranch of Amos Dee Jones from crossing Delaine with Rambouillet, starting in 1920.		M. wt. 125 200 F. wt. 15 22 Grade: Fine, staple	White face, moderately wool-free, as are lower legs. Rather leggy. Good travelers. Some neck folds. Ewes polled, rams either polled or horned.	Debouillet Sheep Breeders' Ass., 300 South Kentucky Ave., Roswell, N.M. 88201
Delaine	Eastern U.S. from imported Spanish Merino.	Original Spanish Merino in 1793.	M. wt. 100–165 150–225 F. wt. 10–12 18–20 Grade: Fine (64's or better)	Moderately tall, deep, but narrow. Some neck skin folds. White face, moderately wool-free. Tight, dense, fine fleece, high in yolk and grease. Ewes polled, rams usually horned. Single lambs generally.	Many recording societies have promoted Merino types. Major current ones are: (1) Texas Delaine Ass., Burnett, Texas 78611 (2) American and Delaine-Merino Record Ass., 4000 Water St., Wheeling, W. Va. 26003
Rambouillet	France, on government experimental farm at Rambouillet, starting in 1786.	1840	M. wt. 150–175 200–250 F. wt. 10–14 20 Grade: Fine to ½ blood	Fairly tall, with legs completely wool covered. White, nearly wool-free face, with some dark spots permitted. Ears, poll, and jaws well wooled and may encroach to cause "wool blindness." Ewes polled, rams either polled or horned. More prolific than other fine-wool breeds.	American Rambouillet Sheep Breeders' Ass., 12709 Sherwood Way, San Angelo, Tex. 76901.

TABLE 4.1. *(continued)*

Breed	Place of Origin	Date Introduced to U.S.	Performance Traits	Distinguishing Characteristics	Official Registration
Fur Karakul	Desert, mountainous area east of Caspian Sea in Russia and Afghanistan, as early as 1400 B.C.	1909	M. wt. 130–160 170–200 F. wt. 6–7 9 Grade: Coarse carpet wool. Lamb pelts in order of value: (1) Broadtail (2) Persian Lamb (3) Krimmer (4) Caracul	Fat-tail type. Fleece tight, curly, long. Face variable solid or mottled colors, wool-free. Head long, narrow; **Roman** nose; ears usually large, long, drooping, but earlessness is acceptable. Ewes polled or semihorned, rams polled or spiral-horned.	Karakul Fur Sheep Registry, Fabius, N.Y. 13063. Certifies flocks, and registers individuals in either Fur Section or Meat Section of Karakul Record.

FIG. *4.1.* (left) *Cheviot ewe.* (right) *Cheviot ram.*
(American Cheviot Sheep Society, Inc.)

FIG. *4.2.* (left) *Columbia ewe.* (right) *Columbia ram.*
(Columbia Sheep Breeders' Assn. of America)

FIG. *4.3.* (left) *Corriedale ewe.* (right) *Corriedale ram.*
(American Corriedale Assn., Inc.)

FIG. *4.4.* (left) *Dorset ewe
(horned).* (upper right)
Dorset ram (horned). (lower right) *Dorset ram
(polled).* (Continental Dorset Club)

FIG. *4.5.* (left) *Hampshire ewe.* (right) *Hampshire ram.*
(American Hampshire Sheep Assn.)

FIG. *4.6.* (left) *Montadale ewe.* (right) *Montadale ram.*
(Montadale Sheep Breeders Assn., Inc.)

FIG. 4.7. (left) *Notail ewe and lamb.* (right) *Notail ram.*
(S. Dak. State Univ.)

FIG. 4.8. (left) *Oxford ewe.* (right) *Oxford ram.* (Larry
E. Mead, Mo.)

FIG. 4.9. *Panama ram.*

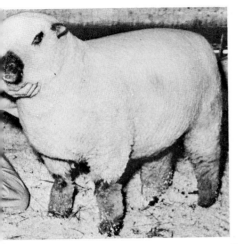

FIG. *4.10.* (left) *Shropshire ewe.* (right) *Shropshire ram.*
(American Shropshire Registry Assn., Inc.)

FIG. *4.11.* (left) *Southdown ewe.* (right) *Southdown ram.*
(American Southdown Breeders' Assn.)

FIG. *4.12.* (left) *Suffolk ewe.* (right) *Suffolk ram.* (National Suffolk Sheep Assn.)

FIG. *4.13.* (left) *Targhee ewe.* (right) *Targhee ram.* (U.S. Targhee Sheep Assn.)

FIG. *4.14.* *Tunis ram.*

FIG. *4.15.* *Cotswold ram.*

FIG. *4.16.* (left) *Lincoln ewe.* (right) *Lincoln ram.*
(National Lincoln Sheep Breeders' Assn.)

FIG. *4.17.* (left) *Romney ewe.* (right) *Romney ram.*
(American Romney Breeders' Assn.)

FIG. *4.18. Debouillet flock on range.* (Debouillet Sheep Breeders' Assn.)

FIG. *4.19.* (left) *Delaine ewe.* (right) *Delaine ram.*
(Texas Delaine Assn.)

FIG. *4.20.* (left) *Rambouillet ewe.* (right) *Rambouillet
ram.* (lower right) *Ram-
bouillet polled ram.* (American Rambouillet Sheep
Breeders Assn.)

FIG. *4.21. Karakul ewe* and *1-week-old lamb.* (Karakul Fur Sheep Registry)

READING LIST

McPhee, H. C. 1936. Breeding problems with sheep. *USDA Yearbook of Agriculture, 1936.* U.S. Government Printing Office, Washington, D.C.
Sheep and general livestock magazines.

[CHAPTER FIVE] Goats [CAPRINES]

GOATS have long been one of man's important domesticated animals and today are found in all countries of the world—the most widely distributed of all domestic animals except the dog. Since prehistoric times they have furnished man with milk; meat; fiber for clothing and housing; skins for bottles, buckets, and clothing; and offerings for Jehovah. In many underdeveloped areas they are still the chief livestock serving human needs. Goats have not, however, been extensively used in modern agriculture.

Both milk goats *(Capra hircus)* and mohair-producing goats *(Capra angorensis)* probably descended from the pasang or Grecian ibex *(Capra hircus aegagrus),* a wild species native to Asia Minor, Persia, and surrounding countries. Only a few of the 60 or more recognized breeds in the world are important in the Americas. The most important mohair-producing breed, the Angora, gets its name from that province in Asia Minor where the principal development of the long-haired type of goats took place. The Saanen, Toggenburg, Nubian, and French Alpine, together with the recently developed American La Mancha, are the principal breeds of milk goats in America. The Maltese breed, important in many areas of the world as a milk goat, has had an influence on the goats in southwestern United States and the Southern Hemisphere. These originated from crosses between the Maltese and Spanish goats and are generally known as Spanish Maltese. Goats of this type and other common American goats of mixed ancestry are used in the Southwest principally for brush control and to a lesser extent for meat. Goat meat is known as chevon.

MOHAIR-PRODUCING TYPES

Mohair-producing goats predate written history. It is known that mohair was in use during the days of Moses and that textiles produced from it were highly regarded.

Mohair was first produced commercially in the province of Ankara in Turkey, which is usually considered the homeland of the Angora. The modern Angora is a larger and hardier animal that produces a heavier and somewhat coarser fleece than the original. This change came about as a result of the high demand for mohair in the early part of the 19th century. The Turks, original breeders of the Angora, were unable to increase their herds rapidly enough to meet the demand. Angora bucks were used in grading up the common Kurd to increase the number of mohair-producing animals in the region. These up-graded Angoras essentially replaced the original purebreds as the foundation of the modern Angora.

While mohair yarn was known in Europe at an earlier date, 1554 marks the first European record of the Angora, when the Dutch Ambassador at Constantinople sent a pair to Emperor Charles V. The breed did not become well established in Europe until the 19th century.

Angora goats were first introduced in South Africa in 1838, and highly meritorious herds were subsequently developed by crossing these with native goats.

James C. Davis of Columbia, S.C., brought the first Angoras from Turkey to the United States in 1849. This importation of seven does and two bucks attracted favorable attention through exhibition at many fairs. The largest early importation—67 head—was made in 1861 by Winthrop W. Chenery of Belmont, Mass., a pioneer importer of many breeds of improved livestock. In 1881 the Sultan of Turkey passed an edict prohibiting the exportation of Angoras, hoping to confine the mohair industry within Asia Minor. By this time, however, the breed was well established in several other countries. One of the most note-worthy importations to the United States consisted of 148 head brought from South Africa in 1904 by G. A. Hoerle of Midland Park, N.J. Temporary suspension of the high export duty of £100 sterling per head during the Boer War made this shipment possible. Soon afterward South Africa imposed an embargo on exportation of Angoras which remained in effect for nearly 20 years. These early export restrictions indicate the worldwide importance of mohair during that era.

U.S. MOHAIR INDUSTRY. Mohair and wool production show similar trends in the United States. Rapid expansion occurred between 1900 and 1930, followed by a declining trend. Angora numbers increased from 329,300 in 1900 to a peak of 3,785,127 in 1930, with over 75% of these concentrated in Texas. The browsing propensity of goats gives them a triple-purpose role—fiber and meat production and control of brush in rangelands. Mohair is used to make special long-wearing cloth for summer suits, knitted outerwear garments, and upholstery and seat covering. Goat meat is similar to lamb and mutton but is of little commercial importance in the United States.

Major wool and mohair scouring plants are located at Brady, Texas, in the Edwards Plateau area, which is a center of both mohair and fine-wool production. Since the middle of the 20th century wool processing, similar to the trend in meat and other animal products, has been shifting from metropolitan manufacturing centers to the production areas. Instead of the traditional shipment of grease wool and mohair to eastern textile mills, the scoured product is now shipped from the concentrated production areas. The oil obtained in scouring is shipped east in tank cars to be used principally in cosmetics for its lanolin content. Scoured mohair weighs about 15–17% less than grease mohair as sheared. Considerable amounts of imported wool from such countries as Argentina and Uruguay are also processed at Texas scouring plants.

PRODUCTION CHARACTERISTICS. The factors of major concern to the producer are yield and quality of mohair and reproductive rate. Unlike milk goats, Angora goats are not highly prolific. Twinning is not common. A well-managed herd may achieve a kid crop close to 100%, but 60–70% is more common in most range herds. The average clip of mohair from all goats in the United States is about 6.5 lb. Pure-bred Angoras may produce twice this amount. The goal in good Texas herds is 12 in. of mohair growth annually, obtained in 6-in. clips at about equal intervals. Angoras are rather delicate at birth and immediately following shearing, and adverse weather changes may occasion considerable loss of animals. Mohair produced during the first year of life has the highest quality and value, mainly due to fineness of fiber. Fiber diameter increases until about the 8th year of life. Males produce somewhat coarser fleeces of shorter fiber length than females.

Genetic improvement efforts have been directed primarily toward fleece characteristics. Elimination of kemp fibers is of particular importance to fleece value, and some increase in oil content helps protect the fibers. Additionally, studies have been conducted on the incidence of cryptorchids and hermaphrodites. These have been fairly common defects in all types of goats. The USDA and state experiment stations in Texas and other producing states, along with producer associations, have provided leadership in improvement efforts.

The breed registry organization for Angoras is the Angora Goat Breeders Association (Rock Springs, Texas 78880), organized in 1900. It was the first such organization in the world for the development and promotion of the breed.

DAIRY GOATS

Dairy goats have for centuries been important contributors of high-quality milk and meat protein in areas of subsistence agriculture. Their small size and efficiency of milk production compared to bovines, together with their prolificacy, give them substantial advantages under such conditions. Goats of this type reached the European continent at an early date, and improved breeds developed there were introduced into the Western Hemisphere. Records indicate that goats of this type were brought to the United States by early settlers in Virginia and in

New England by Capt. John Smith and Lord Delaware. The expansive rather than intensive agricultural trend of the developing new nation gave cattle a higher priority than milk goats for comparable uses.

The first purebred Swiss goats were imported in 1893—a shipment of 4 Toggenburgs. Serious attention to the breeding of milk goats in the United States began shortly after 1900. F. S. Peer of Ithaca, N.Y., one of the noted early importers of Guernsey and Jersey cattle and other improved animal breeds, imported 10 high-quality Saanen and 16 Toggenburg goats in 1904. A number of small importations of both breeds followed during the next 20 years. Nubians, developed in England by crossing bucks from their native northeast Africa with short-haired English does, were brought in during this same period. An importation of 3 bucks and 18 does of the French-Alpine breed was made in 1922 and formed the basis for this breed in the United States. The Maltese, valued as a milk-producing breed in some parts of the world, has had little impact in the United States except as noted earlier in the discussion of mohair-producing goats in the Southwest. The La Mancha, a breed originating in Spain, found its way into California via Mexico following an importation to that country in about 1920. These short-eared goats captured the fancy of a few enthusiastic breeders and formed the basis for the American La Mancha, which was recognized by the American Dairy Goat Association as a separate breed in 1958.

The number of milk goats in the United States has never been large, and they have not been concentrated in a recognizable industry in the same sense as the Angoras. Milk goats have become established throughout the country, mostly adjacent to metropolitan areas where special markets have been developed for goat milk on the basis of its unique physical properties. The composition of goat milk differs little from that of average market milk from cows. However, the fat globules are smaller and a softer curd is formed, thus providing a premium-price market for it in cases where a more easily digested milk is desired. The fat globules are white and do not rise to the top but can be obtained with a separator. Excellent cheese can be made from goat milk. Statistics are not available on the volume of goat milk sold.

DAIRY GOAT IMPROVEMENT. Performance testing for milk and fat production of dairy goats developed along the same lines as that of dairy cows. The 1936 USDA germ-plasm survey elicited little response from private breeders, indicating limited production record-keeping up to that time. Research with dairy goats had been initiated by the USDA in 1909, using purebred bucks of the imported breeds in grading up with native does. Saanen and Toggenburg bucks were used and produced comparable improvement in production. The New Mexico Agricultural Experiment Station started research with dairy goats in 1919, using Toggenburg bucks on native does descended from Spanish introductions into the Southwest. As in the USDA studies, the most marked increases in production were obtained in the first-generation cross, but continued top-crossing resulted in a steady upward trend. Official records at this point were made under the advanced registry rules of the American Milk Goat Record Association. One sire used in the New Mexico herd, Val Verde's Lorenzo 30551 (AMGRA), had both inbred and outbred daughters recorded by 1936. Twenty-two daughters with an inbreeding coefficient of 25.0% averaged 1,306.9 lb. milk and 50.99 lb. fat compared to 1,523.6 lb. milk and 58.23 lb. fat for their dams. His 29 outbred daughters had lactation records averaging 1,624.6 lb. milk and 60.45 lb. fat compared to 1,126.4 lb. milk and 41.21 lb. fat for their dams. The mean gestation length for all does in the herd was 149.9 days, and fertility was high; 286 kids were produced from 144 parturitions, including 4 sets of quadruplets.

At least three different associations have been involved in registering dairy goats. Of these the American Milk Goat Record Association of Vincennes, Ind., organized in 1904, is the oldest, largest, and most influential in inaugurating breed improvement programs. It has continued with the name changed in 1965 to the American Dairy Goat Association (P.O. Box 186, Spindale, N.C. 28160) and registers purebreds, grades, and crosses of all breeds. The other registry organization currently in existence, the American Goat Society, Inc. (1606 Colorado St., Manhattan, Kan. 66502) was initially established in 1935 at Wayland, N.Y., and records only purebreds of the different breeds.

The ADGA established the advanced registry testing program; the AGS initiated the herd improvement registry for complete herd testing in 1936. The evolution of performance testing systems for dairy goats follows the same chronological pattern of AR, HIR, and DHIR testing as that for dairy cattle. They differ in no important details from these and are currently provided for within the framework of modern electronic processing of dairy records, even though still designated as Advanced Registry records.

Improved breeding and feeding methods have resulted in significant advances in recorded performance levels. The 333 Advanced Registry records listed in the 1969 ADGA Handbook for all purebreds and grades averaged 1,948 lb. milk, 3.8% fat, 74 lb. fat in 297 days at an average age of 3 yr., 4 mo. This level of milk production in a lactation approximates 15 times the doe's body weight, which is comparable to the most efficient dairy cows. It is about the same as Holstein cows producing 20,000 lb. milk in a lactation.

Toggenburgs hold the top 305-day lactation records for milk and fat production, both records made in 1960 by half-sisters in the Chikaming herd of Mrs. Carl Sandburg of North Carolina. The top milk record is 5,750 lb. and the top fat record is 202.5 lb. The all-time production leaders, by breeds, as listed in the 1969 ADGA Handbook are shown in Table 5.1.

TABLE 5.1. Top Lactation Performances as of 1971

Breed	Year	Name & Reg. No.	Age (Yr.–Mo.)	Milk (lb.)	Fat (lb.)
French Alpine	1968	7L Silver Bell ☆M A 146891	5—1	4,826	
	1948	Yvon Del-Norte 4☆M A 81191	2—1		171.5
Nubian	1962	Hurricane Acres MacArthur Charade 5☆M N 125061	5—1	4,392	
	1959	Araby Royal Holly ☆M N 121471	3—10		193.9
Saanen	1954	LaSuise Ida-Bee ☆M S 103117	3—0	4,905	
	1962	Morada Pride's Anita 5☆M S 127771	4—1		191.2
Toggenburg	1960	Puritan Jon's Jennifer II 9☆M T 121022	4—0	5,750	
	1960	Puritan Jon's Janista 8☆M T 121020	3—9		202.5
American La Mancha		(No all-time production leaders as yet designated)			

Note: Star in name indicates special recognition based on performance.

VISUAL APPRAISAL. The show-ring provides the principal means of evaluating the type characteristics of dairy goats. Show-ring competition and recognition involve elements common to both dairy cattle and dog shows. The pioneer registry association, now the ADGA, was formed in 1904 mainly to permit the exhibiting of dairy goats at the St. Louis World's Fair. The official score card is patterned almost entirely after the dairy cattle Unified Score Card as approved by the Purebred Dairy Cattle Association and the American Dairy Science Association. Official judges are qualified upon demonstrating their proficiency at scheduled training conferences. The dog show similarity comes about through the winning of Permanent Championship titles upon accumulation of specified credit at ADGA Official Shows. This entitles the designation "Ch." as a suffix to the animal's registered name. In 1968, 117 shows had permits as ADGA Official Shows, most of which were at regularly scheduled livestock fairs or expositions.

ADGA DAIRY GOAT SCORE CARD
(Ideals of type and breed characteristics must be considered in using this card.)
Based on Order of Observation

1. GENERAL APPEARANCE 30
Attractive individuality revealing vigor; femininity with a harmonious blending and correlation of parts; impressive style and attractive carriage; graceful walk.

Breed characteristics   10
Head—medium in length, clean-cut; broad muzzle with large, open nostrils; lean, strong jaw; full, bright eyes; forehead broad between the eyes; ears medium size, alertly carried (except Nubians).
Shoulder blades—set smoothly against the chest wall and withers, forming neat junction with the body.
Back—strong and appearing straight with vertebrae well defined.
Loin—broad, strong, and nearly level.
Rump—long, wide, and nearly level.   8
   Hips—wide, level with back.
   Thurls—wide apart.
   Pin bones—wide apart, lower than hips, well defined.
   Tail head—slightly above and neatly set between pin bones.
   Tail—symmetrical with body.
Legs—wide apart, squarely set, clean-cut, and strong with forelegs straight.
   Hind legs—nearly perpendicular from hock to pastern. When viewed from behind, legs wide apart and nearly straight. Bone flat and flinty; tendons well defined. Pasterns of medium length, strong and springy. Hocks cleanly moulded.   12
Feet—short and straight, with deep heel and level sole.

2. DAIRY CHARACTER   20
Animation, angularity, general openness, and freedom from excess tissue, giving due regard to period of lactation.
Neck—long and lean, blending smoothly into shoulders and brisket, clean-cut throat.
Withers—well defined and wedge-shaped with the dorsal process of the vertebrae rising slightly above the shoulder blades.
Ribs—wide apart; rib bone wide, flat, and long.
Flank—deep, arched, and refined.
Thighs—incurving to flat from the side; apart when viewed from the rear, providing sufficient room for the udder and its attachments.
Skin—fine textured, loose, and pliable. Hair fine.

3. BODY CAPACITY   20
Relatively large in proportion to the size of the animal, providing ample digestive capacity, strength, and vigor.
Barrel—deep, strongly supported; ribs wide apart and well sprung; depth and width tending to increase toward rear of barrel.   12
Heart girth—large, resulting from long, well-sprung foreribs; wide chest floor between the front legs, and fullness at the point of elbow.   8

4. MAMMARY SYSTEM   30
A capacious, strongly attached, well-carried udder of good quality, indicating heavy production and a long period of usefulness.
Udder—Capacity and Shape—long, wide, and capacious; extended well forward; strongly attached.   10
   Rear attachment—high and wide. Halves evenly balanced and symmetrical.   5
   Fore attachment—carried well forward, tightly attached without pocket, blending smoothly into body.   6
   Texture—soft, pliable, and elastic; free of scar tissue; well collapsed after milking.   5
   Teats—uniform, of convenient length and size, cylindrical in shape, free from obstructions, well apart, squarely and properly placed, easy to milk.   4

TOTAL 100

TABLE 5.2 Type Evaluation Summary of Defects (Revised October 1967)

	General	Breed Specifics
Slight	Broken or wry tail	
Slight to serious depending on degree	Undershot or overshot jaw Close in the hocks Front, rear, or side udder attachment lacking Separation between halves of udder or presence of scar tissue Udder of beefy texture or with pocket	

Moderate	Large scurs or stubs
	Enlarged knees; nondisabling lameness
	Turned-out or crooked feet
	Teats:
	  Set close together
	  Bulbous
	  Extremely large or small
	  Pointed sideways
	  Uneven in size
	  Having small streams or otherwise hard to milk
	  Not clearly separated from the udder
		American La Mancha
		  Mature does less than
		    minimum height (28 in.)
		    minimum weight (130 lb.)
		French Alpine
		  Mature does less than
		    minimum height (30 in.)
		    minimum weight (135 lb.)
		  Does with Toggenburg color
		    and markings
		Nubian
		  Mature does less than
		    minimum height (30 in.)
		    minimum weight (135 lb.)
		  Straight face
		Saanen
		  Mature does less than
		    minimum height (30 in.)
		    minimum weight (135 lb.)
		Toggenburg
		  Mature does less than
		    minimum height (26 in.)
		    minimum weight (120 lb.)
		  Few small white spots in hair
		    of does

Moderate to serious depending on degree	Loose, winged, or heavy shoulders
	Narrow chest or pinched heart girth
	Short, shallow, or narrow body
	Low-backed or steep-rumped
	Small-boned for body size
	Bowed-over front knees, buck-kneed
	Hind legs close together
	Sprung pasterns
	(All of these more serious in bucks)
		American La Mancha
		  Roman nose
		French Alpine
		  Roman nose
		Saanen
		  Roman nose
		Toggenburg
		  Roman nose

Serious	Natural horns (Neatly disbudded or dehorned—no discrimination)
	Udder:
	  Pendulous
	  Too distended to determine texture
	  Hard or swollen (except in does just fresh)
	  So uneven that one half is less than half the size of other
		French Alpine
		  Bucks with Toggenburg color
		    and markings
		Saanen
		  Dark cream color
		  Several small dark spots in hair
		Toggenburg
		  Black color in does
		  White stomach (except British Toggenburgs) on does

TABLE 5.2. *(continued)*

	General	Breed Specifics
Serious *(cont.)*	Leaking orifice	Large white spot (1½″ or more in any direction) on does Few small white spots in hair of bucks

Very serious	Udder lacking in size and capacity in relation to size of doe
	Double orifice in teat
	Extra teat or teats that have been cut off on does
	Crooked face on does
	Very crooked or malformed feet
		Nubian
		  Dished face
		  Barely drooping ears

Disqualifications	Total blindness
	Serious emaciation
	Permanent lameness or difficulty in walking
	Blind or nonfunctioning half of udder
	Blind teat
	Double teats
	Extra teats that interfere with milking
	Active mastitis or any other cause of abnormal milk
	Evidence of hermaphroditism or other inability to reproduce
	Permanent physical defect, such as navel hernia
	Crooked face on bucks
	Extra teats or teats that have been cut off on bucks
	Double orifice in teats of bucks
	Buck with one testicle or with abnormal testicles
		American La Mancha
		  Anything other than gopher ears on bucks
		  Ears other than true La-Mancha type on does
		French Alpine
		  Pendulous ears
		Nubian
		  Upright ears
		Saanen
		  Large (1½″ diameter or more) dark spot in hair
		  Pendulous ears
		Toggenburg
		  Tricolor or piebald
		  White stomach (except British Toggenburgs) on bucks
		  Large white spot (1½″ in any direction) on bucks
		  Pendulous ears

PROMOTION ACTIVITIES. In addition to shows and consignment sales, the major breeds have active national breed associations devoted to promotion activities. Publications of general interest include the annual ADGA Handbook, which contains advertising, general information, and official reports of shows and production records. The *Dairy Goat Journal* (Box 836, Columbia, Mo. 65201), published monthly, is an official publication of the American Dairy Goat Association.

Comparative numbers are indicated by the number registered for each breed by the ADGA in 1969:

Nubian	2,531
French Alpine	1,357
Toggenburg	869
Saanen	787
American La Mancha	160

TABLE 5.3. Summary of Goat Breeds

Breed	Size				Body Conformation	Color	Distinguishing Characteristics
	Females		Males				
	Height (in.)	Weight (lb.)	Height (in.)	Weight (lb.)			
Mohair							
Angora		75–120		125–175	Small; straight top and bottom lines; blocky; narrow; moderately tall; light bone; light muscling throughout; small udder.	Pure white; kids sometimes red at birth; blacks may occur but are not registered.	Head medium to small; straight profile; refined. Both sexes have horns (small in females, spiraled and up to 2 ft. in length in males). Ears long, heavy, and droopy; mostly covered with hair. Coat hangs in ringlets or small flat curves.
Dairy							
American La Mancha	28	130	33	170	See ADGA Score Card	Multicolored, with whites and animals with white marks predominating. Hair coat smooth and fine.	Head medium in size, relatively wide; straight profile. Horns may occur but most individuals are hornless. Both sexes may have beards and/or wattles, but latter are often removed. Ears short and stubby.
French Alpine	30–40	135	34–40	180	See ADGA Score Card	Multicolored, with variations from pure white through shades of fawn, gray, brown, black, red, buff, piebald, or combinations. Hair coat smooth and short with males having long, coarse hair along spine.	Head medium to long; profile straight; eyes set well back in head. Both sexes may be polled or horned and have beards and/or wattles. Ears medium to small, refined, carried erect. Average production highest of all breeds in U.S.

TABLE 5.3. (continued)

Breed	Size				Body Conformation	Color	Distinguishing Characteristics
	Females		Males				
	Height (in.)	Weight (lb.)	Height (in.)	Weight (lb.)			
Nubian	30	135	36	175	See ADGA Score Card	All colors or combinations of colors; black, red, tan, and combinations of these with and without spots most common. Hair coat short and smooth in females, longer and coarser in males.	Head large, broad, with a distinct Roman nose; eyes prominent; nostrils deeply depressed. Mostly hornless but horned individuals exist. No beards; may or may not have wattles. Ears large, drooping downward and forward.
Saanen	30–36	135	35–40	185	See ADGA Score Card	White or cream. Hair coat short and smooth.	Head medium in size, very refined; face straight or slightly dished. Both horned and polled individuals exist. Both sexes have beards and may have wattles. Ears medium in size, refined, carried erect.
Toggenburg	26–32	120	33–38	160	See ADGA Score Card	Solid colored body varying from light brown to dark chocolate; two white stripes on face; white on outside of ears, on wattles or wattle area, on legs below knees and hocks, and around tail. Hair coat short and soft on females, coarse and longer on males.	Head short and moderately broad; profile straight. Mostly hornless but horned individuals of both sexes exist. Both sexes have beards and may have wattles. Ears medium in size, refined, carried erect.

FIG. 5.1. (left) *Mature Angora buck.* (right) *Champion Angora kids.*

FIG. 5.3. *French Alpine doe.* (Mr. and Mrs. Donovan A. Beal, Turlock, Calif.)

FIG. 5.2 *American La Mancha doe.* (R. W. Soens, Bostic, N.C.)

FIG. 5.4. *Nubian buck.* (American Dairy Goat Assn.)

FIG. 5.5. Saanen doe.
(American Dairy Goat
Assn.)

FIG. 5.6. Toggenburg doe.
(W. and D. Linder, Calif.)

READING LIST

ADGA Handbook (annual volumes). American Dairy Goat Association, Spindale, N.C. 28160.

Dairy Goat Journal (monthly magazine). Columbia, Mo.

French, M. H. 1970. *Observations on the Goat.* FAO Agricultural Studies No. 80, Rome.

Lambert, W. V. 1937. Breeding problems with Angora goats. *USDA Yearbook of Agriculture, 1937.* U.S. Government Printing Office, Washington, D.C.

Simmons, V. L., and W. V. Lambert. 1937. Improvement of milk goats. *USDA Yearbook of Agriculture, 1937.* U.S. Government Printing Office, Washington, D.C.

[CHAPTER SIX] Horses, Ponies, and Asses [EQUINES]

EQUINES are used in all countries of the world. Viewed chronologically and geographically they have served nearly every use made of domestic animals: transportation, agricultural and industrial power, food (meat and milk), leather goods, and recreation. The horse holds somewhat the same position in the Western Hemisphere as the sacred cow holds in India, with strong taboos against human consumption of its flesh.

The horse (including ponies), donkey (ass), and zebra belong to the suborder Hippomorpha within the order Perissodactyla (uneven-toed). They are usually referred to as belonging to the Equidae family. Horses and ponies are *Equus caballus,* asses *Equus asinus,* and zebras *Equus burchelli.* All these will interbreed. The zebra has never been domesticated, but individuals have been tamed to harness and many are found in zoos. The mule, sometimes referred to as the farm animal "without pride of ancestry or hope for posterity," is the result of crossing a jack (male ass) and a mare. The reciprocal cross (stallion x jennet) is called a hinny and has seldom been made, probably because of the greater utility of brood mares than of jennets as well as the reluctance of stallions to mate with jennets. Virtually 100% of the hybrids in either case are sterile.

HORSES AND PONIES

Since the distinction between horses and ponies is essentially a matter of size, they can be spoken of collectively. The major subdivisions are (1) draft horses, (2) light horses, and (3) ponies, with discontinuity between adjoining groups being somewhat arbitrary. The evolution of the horse has been traced more completely than perhaps any other species, with the conclusion that the distant progenitor was a small,

[103]

doglike animal less than 12 in. tall called the Eohippus—or dawn horse—of the Eocene period. Our discussion will take a long jump forward in time to the more recent origins of equines that have become important in the Western Hemisphere.

It is rather generally accepted that the Arabian breed holds the record for antiquity among our improved breeds of livestock, probably descending from the wild Libyan horse of northern Africa. Their domestication in Egypt presumably took place several centuries before the Christian era, with acquisition and development by the Arabs beginning sometime between the 1st and 6th centuries. The Arabian or its close relatives have apparently contributed to all of the improved breeds of horses subsequently developed.

Viewed from the standpoint of horses and ponies important in the Americas today, horse breeding emphasis has come full circle—from the mobile, durable, and swift "horse of the desert" through eras of draft horse and miniature pony importance back to types among which the Arabian characterizes the intermediate. Spanish horses of this type were the first to reach western shores, brought over in substantial numbers by the Spanish explorers in the 1500s. Many of these escaped to the wild where they thrived and multiplied to form the wild herds of mustangs or broncos of the western ranges. The utility of these animals for both the Indians and cowboys is amply documented in the lore and legend of the West. Later introductions by the colonists to the Atlantic coast included many improved breeds of both draft and light horses. These provided essential services to the developing nations of the Western Hemisphere—power, transportation, communication, recreation, military strength—until mechanization replaced them for many uses. Except for stock handling on the range and farm use in a few less developed agricultural areas, today's use of the horse is largely recreational.

NUMBERS AND DISTRIBUTION. Estimates of world horse numbers showed over 75 million in 1953 and 69 million in 1960, the last year in which domestic or world numbers were reported by the USDA. The leading countries in horse numbers during this period were (1) USSR, (2) Brazil, (3) China, (4) Argentina, (5) Mexico, (6) United States, (7) Poland, (8) France, (9) India, and (10) Colombia. Horse numbers for utilitarian purposes have declined most rapidly in recent years in those countries where modernization with tractor, truck, and automobile power is most nearly complete. Conversely, a resurgence of horse numbers for recreational purposes has taken place in the more affluent and advanced countries such as the United States and Canada.

Horses and mules reached their peak numbers in the United States in the early 1920s following World War I with over 25 million on farms and 2 million in cities. Numbers had declined to slightly over 3 million in 1960. The number of draft horses and mules is now insignificant, but light horses and ponies are increasing rapidly. No reliable estimates of current total numbers are available, but indications are that they are well over twice the 1960 figure. This is suggested by a 96% increase in registrations of purebred horses between 1960 and 1968.

According to the 1960 count, the leading states in horse numbers were (1) Texas, (2) North Carolina, (3) Kentucky, (4) Mississippi, (5) Tennessee, (6) Alabama, (7) Georgia, (8) Missouri, (9) Louisiana, and (10) Virginia.

THE HORSE INDUSTRY. The passing of the draft horse, except in the hands of a few hobbyists and some well-known exhibition hitches such as those of the Anheuser-Busch brewery firm and Wilson & Co. meat packers, is viewed with some nostalgia by elderly Americans who have experienced the giant step from the horse-and-buggy era to the modern space age. But as a major factor in the economy the draft horse is a thing of the past, and hardly anyone would wish to turn back the clock to the walking plow or the "surrey with the fringe on top." For the aspiring young animal agriculturist to whom hames, breeching, and traces are unfamiliar livestock terms, the demise of the draft horse as a utilitarian animal provides a valuable lesson. L. B. Wescott, a leading New Jersey breeder of registered Suffolk horses and Guernsey cattle, stated the case for utility in a 1946 editorial: "The $5,000 brood mare was never expected to pay for herself hauling manure; but when the market for horses to haul manure disappeared, the $5,000 brood mare went with them, and no amount of money or blind devotion to blooded stock made one particle of difference." Whether the need be food, fiber, companionship, or recreation, a breed or species soon ceases to be a viable part of the animal industry if it fails to compete in meeting that need. The rank-and-file dairy bull is no longer a fixture on commercial dairy farms, because artificial insemination service to superior bulls eliminated the need for him. No amount of merchandizing expenditure or blind devotion to maintaining extensive bull sales made a particle of difference here either.

The horse industry has again thrived by virtue of the light horses and ponies that meet a recreational or sporting need of an affluent society with leisure time. Over the past decade the number of official horse shows held under auspices of the American Horse Shows Association has doubled and accounts for nearly one-third of the total of some 3,000 horse shows held annually in North America. Nearly 100 polo clubs are in operation, and there are over 100,000 miles of riding trails in 154 national forests and 19 national grasslands. In 1970 there were 225,000 horse projects in 4-H clubs. Although not acceptable for registered purebreds, the recent achievement of successfully freezing horse semen and making top sires available through AI should facilitate the acquisition of better horses for junior project members and other horse owners.

Horse racing, one of the nation's largest spectator sports, drew 66

FIG. *6.1.* (above) *Famous 8-horse hitch of Clydesdales.* (Anheuser-Busch, Inc.) (left) *Hackney pony in action.* (American Hackney Horse Society)

FIG. *6.2. Famous Standardbred pacer, Bret Hanover, by Adios.* (U.S. Trotting Assn.)

FIG. *6.3. Standardbred trotting mare, Rosalind.* (U.S. Trotting Assn.)

FIG. 6.4. *Connemara Pony mare, Whitewood Muffin, shown as English pleasure mount. Note English-type equipment and riding habit.* (American Connemara Pony Society)

FIG. 6.5. *Paint stallion, Snip Bar, shown as Western pleasure or working mount. Note Western-type equipment and riding garb.* (American Paint Horse Assn.)

million people to 230 tracks in 1970. Admittedly a high percentage of those attending come to wager money rather than for love of the horse. Thirty states permit pari-mutuel betting, which returns large revenues to the state treasuries. The average bettor gets a return of something like 85 cents per dollar wagered; the balance is used to pay track expenses, purses, and the state's share of the returns. Some $5 billion are wagered annually, and another billion is pumped into the economy incidental to racing operations—for feed, equipment, training, transportation, lodging, and food.

USE CLASSIFICATION. The three major subdivisions (draft, light, and pony) have already been noted. Light horses and ponies, which constitute the most important groups, can be further classed as:

1. Carriage or heavy harness horses
2. Roadster or light harness horses
3. Fine harness horses
4. Riding horses
   a. Runners
   b. Hunters and jumpers
   c. Polo
   d. Western
   e. Parade
   f. English
   g. Plantation walking
   h. Cavalry mounts
5. Ponies (some of the same use classes apply here as for 3 and 4 above)

Some factors that affect use classification of members of the horse family are size, conformation, speed, gait or "manner of going," and temperament. Size is expressed by height at the withers and by weight. In horseman's parlance, the unit of height is the *hand* (1 hand = 4 in.). A horse's height is given in hands and inches (e.g., a horse that measures 62 in. at the withers would be listed as 15-2, meaning he is 15 hands and 2 in. tall. The height of ponies is usually expressed directly in inches, with a maximum height to qualify as a pony usually specified. The performance traits are affected by genetic capability and training to varying degrees. A particular breed or type of horse may be adaptable to training for several different uses, such as a combination riding and light harness horse or as either a runner or a hunter and jumper. On the other hand, the inherent high head carriage, leggy conformation, and spirited temperament of a typical English-type mount would make it a very unhandy kind of horse to use in roping cattle, regardless of training.

EVALUATING PERFORMANCE. Because of the uses made of horses, soundness and temperament are the most important factors in their evaluation. A high degree of utility and versatility can be achieved by horses exhibiting a wide range of types, sizes, and degrees of conformation excellence if they are sound and tractable.

During the era of major draft horse importance, pulling tests were developed to measure their ability. The dynamometer was developed to record the exact tractive power exerted by a team. Tractive pull of horses in good working condition is related to body weight, showing a correlation of about 0.6 according to dynamometer tests. Horse pulling contests, similar to today's tractor pulling contests, have been popular events at agricultural fairs, using a dynamometer or loaded sled. Some hobbyists maintain draft horses today for competition in such contests.

The preponderance of performance evaluation today with light horses and ponies centers around sporting or recreation events. Those involving the largest number of horse owners are trail rides, amateur rodeos, and performance classes at shows. Many breeds are used in these events, by far the largest number of which fall into the general category of Western pleasure or working horses. A competitive adaptation of the trail ride is the endurance ride, which has been rather widely used and publicized by some breeds. Many agricultural colleges, especially in the western half of the country, have student rodeo clubs and teams which compete in intercollegiate competition. Four-H Club competition in junior horse and horsemanship shows has enjoyed a tremendous boom since 1960. These include showing at the halter, equitation or performance classes, and some junior rodeo events.

Racing is the big money segment of the horse business, highly specialized and professional. The classic racing circuit is the domain of the Thoroughbred, and Kentucky has long been famed as the breeding ground of many of the racing greats. Speed is the all-important criterion, usually at distances of 1–1¼ miles. (The distance may be expressed in furlongs which are ⅛ mi. or 220 yd.) Perhaps the most coveted racing achievement in America is the winning of the "triple crown," consisting of the Kentucky Derby, the Preakness, and the Belmont Stakes for 3-year-olds. No horse has won the crown in recent years, although in 1971 Canonero II, a Kentucky-bred, Venezuelan-owned colt won the first two races before finishing fourth in the Belmont Stakes. High prices are paid for stallions that have posted important wins and fast track times, and their services as sires are much sought after. In 1970, for the first time since 1962, Hail to Reason displaced Bold Ruler as the leading sire of money winners. The highest price recorded to date for a Thoroughbred was $5,440,000 paid in 1970 for the Canadian-bred, English "triple-crown" winner, Nijinsky II. Racing performance, even though heavily influenced by the skill of trainer and jockey, is the major criterion of value and breeding emphasis.

Following are some statistics on Thoroughbred racing for 1970:

Total purses	$182,313,787
No. of races	54,925
Av. purse per race	$3,319
No. of runners	47,769
Av. earnings per runner	$3,817
No. of sires represented among the 840 North American stakes winners	540
No. of yearlings sold at auction	3,166
Av. price per yearling sold	$7,799
(From *Thoroughbred Record*, Feb. 20, 1971)	

The Thoroughbred breed has contributed to the development of many other breeds, including the Standardbred and the Quarter Horse.

The Standardbred was developed principally to produce trotters and pacers for harness racing, another sporting event enjoying long-time popularity. In addition to pari-mutuel tracks, harness racing was at one time as popular at county and state fairs as automobile races are today. The pace, in which the front and rear leg on the same side rather than the diagonals move together, is a faster gait than the trot. Trotters have generally been preferred, however, and the breeding of pacers is somewhat incidental to the production of trotters. Hambletonian 10, foaled in 1849 in Orange County, N.Y., was the most famous sire of trotters; the classic trotting race for 3-year-olds bears his name. The Hambletonian Stake was run for many years at Goshen, N.Y., but is now held at DuQuoin, Ill. The fastest times for 1 mile for both trotters and pacers has been slightly under 2 minutes since just before the turn of the century. As a contributor to human welfare it is worthy of note that many of today's older generation owe their safe arrival into the world, in isolated farm homes, to the mobility of the family doctor at the reins of a good buggy-horse typified by the Standardbred. What truer measure of performance could man ask of animals that have served his needs?

The American Quarter Horse, so named because of its speed at distances of about a quarter-mile, is increasingly reaching into the racing field. Quarter Horse racing was conducted at 103 tracks in the United States and Canada in 1970, of which 72 were pari-mutuel and 31 non-wagering tracks. Purses totaled nearly $9.5 million, an increase of nearly $1 million over 1969. The American Quarter Horse Association sponsors a Register of Merit recognition for racing performance, in which a Speed Index was introduced in 1970 as the basis for qualification. The leading money-winner at 440 yd. and under in 1970 was Rocket Wrangler, whose winnings of over $250,000 in 13 starts were nearly twice those of his nearest competitor.

The Quarter Horse is especially popular with 4-H horse project members. An American Junior Quarter Horse Association was established in 1970 to further these activities. Quarter Horse registrations totaled 90,877 in 1970 for a 37% increase over 1969, and total registrations through 1970 exceed the combined total of all other breeds. The breed is still in the formative stage and utilizes an open studbook procedure to accept new blood into the breed, especially the progeny of outstanding Thoroughbred sires.

Many other performance capabilities of horses are measured in a variety of competitions. Calf roping, barrel races, steer wrestling, and other rodeo events measure the ability of both horse and rider. Hunters and jumpers compete in steeple chases and specified indoor obstacle courses at horse shows. Dressage represents a specialized form of equitation and is a feature in Olympic competition. The ability of horses to perform at particular gaits is an important part of the evaluation in many show-ring classes, such as the three- and five-gaited classes featuring the American Saddle Horse, the walking gait (plantation walking) of the Tennessee Walking Horse, the high knee action of the Hackneys, and the smooth, short-striding pace of the Peruvian Pasos. The basic natural gaits of the horse are the walk, trot, and canter. The canter is essentially a gallop performed at restrained speed and in shorter stride. Additional gaits which many horses can be trained to perform—coming more naturally to some breeds than to others—are the pace, slow gait (running walk), fox-trot, and rack. The running walk is a specialty of the Tennessee Walking Horse. The fox-trot is a broken trot in which the hind foot reaches the ground slightly ahead of the diagonal forefoot. The rack, formerly more descriptively termed "single-foot," is a rapid four-beat gait that is showy and comfortable for the rider but exhausting for the horse.

HORSE SHOWS. The element of visual appraisal of form and function enters strongly into horse shows. This is the essence of halter classes for all types of horses. However, unlike shows of other classes of livestock, a horse show program features many classes which provide entertainment for general audiences. The costume class for Arabians, the riding and driving classes, the three- and five-gaited classes, and the impressive six-horse Clydesdale hitches are examples of crowd-pleasing events that appeal to the general public.

The ability to generate gate receipts helps to account for the continued growth of horse shows, in contrast to the financial stress experienced in recent years by other livestock exhibitions. The horse industry has exerted major efforts to establish and enforce high standards of ethics in the conduct of horse shows. The American Horse Shows Association (527 Madison Ave., New York, N.Y. 10022) is the major organization representing show interests. It licenses qualified judges and specifies the rules for shows held under its auspices, and its influence has a favorable effect in upgrading the standards of the many additional shows not encompassed by its official sanction.

A complete discussion of shows and show classes is too lengthy for inclusion here. Textbooks covering breed standards and score cards, unsoundnesses of horses, and colors and markings are readily available. The literature on horses is perhaps more copious than that of any other species, with well over 2,000 books on the horse in the English language.

OTHER CONTRIBUTIONS OF THE HORSE. The invaluable role of the horse in the development of this and many other nations elicited heavy emphasis by the veterinary profession on its anatomy and physiology. This knowledge carried over into both human medicine and the treatment of other classes of livestock. Horses have also been used for production of vaccines such as tetanus antitoxin and pneumonia antiserum. Pregnant mare serum (PMS), a follicle-stimulating substance obtained from the blood serum of mares 60–80 days pregnant, has been prescribed by physicians in treating fertility problems. It is also important in new research on estrous cycle synchronization and multiple ovulation in farm animals.

ASSES

This member of the equine family, presumably descended from the wild asses of Asia and Africa, has little present-day importance in North America. When draft horses and mules were important for agricultural, industrial, and military uses, the American Jack was an influential breed in the production of mules.

The small donkey is still a very important factor in the daily life of many of the less privileged people of Central and South America. The significance of this reliable little animal is readily apparent by the number seen carrying packs, pulling loads, or being ridden along the roads of Colombia, Venezuela, Ecuador, and other Latin American countries. The ass has served as the "poor man's horse" in many areas of the world since biblical times. His sure-footedness, docility, endurance, and ability to withstand the rigors of nutritional and environmental stress have made him a valuable contributor to human needs.

Selective breeding and breeds have been little emphasized except during the era of extensive mule production, when improvement of the American Jack received considerable attention. Missouri, Kentucky, and Tennessee were major centers of this effort. George Washington was the first American breeder of Jacks, starting with introductions in 1787, and is known to have encouraged the production of mules for draft purposes.

The miniature donkey, similar to South America's utility animal, is occasionally found in North America as a children's pet or hobby animal.

TABLE 6.1. Summary of Equine Breeds

Breed	Place of Origin	Date Introduced to U.S.	Performance Traits	Distinguishing Characteristics	Official Registration*
Light Horses (Imported, riding) Arabian	Arabia in SW Asia	1765 Ranger, to Connecticut, sire of George Washington's gray Revolutionary War mount.	† Ht.: 14-3 to 15-0 Wt.: 850–1,000 Used mostly for pleasure and exhibition riding, with either western or English equipment.	Variable solid colors with or without white markings. Broken or spotted colors objectionable. Characteristic "chiseled" face, well-arched neck, short back.	Arabian Horse Club Registry of America, Inc., 7801 East Belleview Ave., Englewood, Colo. 80110. Formed in 1908. Closed studbook.
Thoroughbred	England. Raced there as early as 1377.	1730	Ht.: 15-0 to 16-2 Wt.: 1,000 Used for racing under saddle. Contributed to many other breeds of light horses. Good speed for the mile is about 1:35.	Color not important, but bay most common in U.S. Speed rather than conformation is criterion of value. Long, easy, powerful stride; well-muscled rear quarters.	Jockey Club, 300 Park Ave., New York, N.Y. 10022. Formed in 1894. Closed studbook. Natural service. Identity by color sketch, photo, and lip tattoo. Club also sets racing regulations and licenses trainers and jockeys.
Light Horses (American, riding) American Saddle Horse	Developed in southeastern U.S. soon after 1800, from Thoroughbred, Standardbred, Arabian, Morgan, and native stock descended in part from Canadian pacers.		Ht.: 15-0 to 16-2 Wt.: 1,000–1,150 Used mostly for pleasure and exhibition riding with English equipment. Excels in 3- and 5-gaited competition.	All normal dark solid colors, with white markings desirable. Chestnut, bay, brown, and black preferred. Spirited, high head carriage, long slender neck. Show horses usually have tails "set."	American Saddle Horse Breeders Ass., Inc., 929 So. 4th St., Louisville, Ky. 40203. Formed in 1891, under slightly different name. Both closed and open studbook entries. Chain of title for registered stock required.

* The age of a horse is reckoned in years as of January 1 of each year following its birth. Example: A foal born anytime during calendar 1970 would be considered 1 yr. old on Jan. 1, 1971.
† The height of an equine is measured in hands and inches (e.g. 14-3 = 14 hands, 3 inches).

TABLE 6.1. *(continued)*

Breed	Place of Origin	Date Introduced to U.S.	Performance Traits	Distinguishing Characteristics	Official Registration*
American Albino	Started in 1918 at C. R. & H. B. Thompson's White Horse Ranch, Naper, Neb., with Morgan mares bred to all-white stallion, Old King, sire of many circus horses.		Ht.: 14-0 to 16-0 Wt.: 1,000–1,200 Used mostly for exhibition, parade, circus, or trick horses.	All-white with pink skin. Not a true albino, since eyes and extremities contain pigmentation. Typical saddle horse conformation.	American Albino Ass., Inc., Crabtree, Ore. 97335. Formed at Naper, Neb., in 1937. Open studbook. Several pigmentation types of albino, and off-color types recorded.
Appaloosa	Developed by Nez Perce Indians in Palouse River area of Oregon and Washington, starting about 1730, from early Spanish introductions to Mexico. Distinctive color pattern is of ancient origin.		Ht.: 14-2 to 15-3 Wt.: 950–1,100 Typical western pleasure or stock horse conformation and utility. Must be at least 14-0 tall for registry.	Unique spotted color pattern, especially over loin and croup. Eye has white area like humans, skin parti-colored, hoofs black with white vertical stripes. Variable degrees of basic colors and white with typical colored spots are acceptable. No Albino, Pinto, or Paint breeding or color markings.	Appaloosa Horse Club, Moscow, Idaho 83843. Formed in 1938. Studbook provides both closed book foundation registry and open book provisions. A rapidly growing breed, ranking only below Quarter Horse and Thoroughbred in current annual registrations.
Morgan	Developed in Vermont from the foundation sire, Justin Morgan, foaled in 1789 of uncertain ancestry. Probably Arabian and Thoroughbred breeding predominated in his background.		Ht.: 14-2 to 15-2 Wt.: 1,000–1,200 All-purpose riding or light harness horse. USDA and Vt. Agr. Exp. Sta. maintained a Morgan Horse Farm at Middlebury, Vt., from 1905 until mid-1940s. Endurance tests developed by USDA used for selection. Breed excels in trail rides. A rigid Justin Morgan Performance Class sponsored at some shows.	All normal horse colors exist, the dark ones being most popular. Rather compact, smooth, rounded conformation. Alert, intelligent, even temperament, without nervousness.	Morgan Horse Club, Inc., P.O. Box 2157, Bishop's Corner Branch, West Hartford, Conn. 06117. Formed in 1909. Registry rules have varied. Since 1948, has functioned strictly as closed studbook.

TABLE 6.1. (continued)

Breed	Place of Origin	Date Introduced to U.S.	Performance Traits	Distinguishing Characteristics	Official Registration*
Paint (May be double-registered as Pinto.)	Southwestern U.S., from early Spanish introductions. Original type much prized by Indians and early cowboys. Quarter Horse and Thoroughbred breeding introduced recently.		Ht.: 14-0 or taller Wt.: 900–1,100 Only the three natural gaits desired, the pace and rack not permissible. All-purpose riding type. Shown under either western or English equipment.	Dark or light splashes on contrasting body color. Two basic color patterns are: (1) Tobiano—dark splashing on white body. (2) Overo—white splashing on dark body color. General color definitions would be piebald (white and black) and skewbald (white and some dark color other than black).	American Paint Horse Ass., P.O. Box 12487, Fort Worth, Tex. 76116. Formed in 1962 under slightly different name. Open studbook, with registry under 3 classifications. Color and conformation more important than ancestry.
Pinto (May be double-registered as Paint.)	Southwestern U.S. from early Spanish introductions prized mainly by Indians. Color pattern, like Paint, is of antique origin. Closely related to Paint, but generally smaller.		Ht.: 12-0 to 16-0 Wt.: 750–1,100 About 15-0 and 1,000 lb. preferred. No discrimination against pacers. All-purpose riding type. Often shown in same class with Paints.	Same colors as Paint.	Pinto Horse Ass. of America, Inc., P.O. Box 3984, San Diego, Calif. 92103. Formed in 1956. Registry requirements similar to Paint, open studbook.
Palomino	Early 1900s by selecting desired color, using Arabian, Quarter Horse, American Saddle Horse, Thoroughbred, Morgan, Tennessee Walking Horse, and Standardbred stock.		Ht.: 14-2 to 16-0 Wt.: 900–1,200 Main use is for exhibition or parade mounts with fancy western equipment. Good all-purpose riding horse, but color is major criterion of selection.	Typical palomino color (golden chestnut with flaxen mane and tail). May have white markings on face and legs. Not a true-breeding color, but similar to roan Shorthorn color inheritance. Chestnut X albino mating appears to give only Palomino foals.	Palomino Horse Breeders of America, P.O. Box 249, Mineral Wells, Tex. 76067. Formed in 1941. Two other organizations also register Palominos, and some purebreds of other breeds exhibiting this color are double-registered. Open studbook based on color, with different registry categories.

TABLE 6.1. *(continued)*

Breed	Place of Origin	Date Introduced to U.S.	Performance Traits	Distinguishing Characteristics	Official Registration*
Quarter Horse	Mainly in southwestern U.S. based on foundation of Arab, Turk, and Barb types introduced by early Spanish explorers, and developed as stock horses. Crosses of these with Thoroughbred, Morgan, Arabian, and Standardbred produced first registered Quarter Horses.		Ht.: 14-2 to 15-2 Wt.: 950–1,200 Highly versatile all-purpose riding horse, but developed chiefly for working livestock and short-distance racing. Shown under western equipment. Very tractable and reliable temperament. Performance is emphasized in racing and working horse competition.	All normal horse colors except Appaloosa, albino, or spotted acceptable. Darker colors preferred. Intermediate, well-muscled type. Rather low head carriage.	American Quarter Horse Ass., P.O. Box 200, Amarillo, Tex. 79105. Formed in 1940. Several other registry organizations record horses of this type. Open studbook, divided principally into (1) Numbered and (2) Tentative recording. Leading breed in number of registrations.
Tennessee Walking Horse	Developed mainly in Tennessee and surrounding southeastern area, originally known as Plantation Walking Horse. Mixed ancestry includes Canadian Pacer breeding.		Ht.: 15-0 to 16-2 Wt.: 1,000–1,200 Natural gaits are walk, running walk, and canter, with major points assigned to each in judging. Rated as outstanding pleasure riding horse. Famous for the running walk, which provides an easy ride at 6–8 mi. per hour. Usually shown under English equipment.	All normal horse colors exist, frequently with white on face and legs. Head not carried high when moving, and shows characteristic nodding and swinging of the ears at the running walk. Gaits are easy for both horse and rider. Tails of show horses often "set."	Tennessee Walking Horse Breeders & Exhibitors Ass. of America, P.O. Box 87, Lewisburg, Tenn. 37091. Formed in 1935. Closed studbook. Animals resulting from AI not acceptable for registry as a breeding animal.

TABLE 6.1. *(continued)*

Breed	Place of Origin	Date Introduced to U.S.	Performance Traits	Distinguishing Characteristics	Official Registration*
Light Horses *(Light harness)* Standardbred (or American Trotter)	Eastern U.S., mostly in the 1800s, originally as a utility buggy horse. Hambletonian 10, great-grandson of imported Thoroughbred, Messenger, was most influential foundation sire. Norfolk Trotter, Canadian Pacer, Morgan, Arabian, and Barb stock also included.		Ht.: 14-2 to 16-2 Wt.: 800–1,250 Major criterion of selection has been speed in harness at trot or pace. One-mile time of under 2 min. is common at either gait for best horses, and is most common distance for U.S. races.	Bay most common, but other normal horse colors occur. Inherent ability to trot or pace at high speed is chief distinction.	United States Trotting Ass., 750 Michigan Ave., Columbus, Ohio 43205. Formed in 1871, under a different name. **Registers** Standardbreds in both U.S. & Canada. Detailed registry requirements under several conditions, mostly of closed book concept, although open book procedure is provided for. Association also sponsors and governs racing procedures.
Light Horses *(Heavy harness)* Hackney (at one time called Norfolk Trotters in England)	England	1822	Ht.: 15-0 to 16-0 (Classed as ponies if under 14-2) Wt.: variable, depending on height. Society shows usually divide them according to height. An elite carriage horse prior to the automobile. Now used only for exhibition. Noted for high, exaggerated knee action.	Bay most common in U.S., but any normal dark horse color, with or without limited white markings, acceptable. Powerful action along with speed and endurance at the trot.	American Hackney Horse Society, 527 Madison Ave., Room 725, New York, N.Y. 10022. Closed studbook.

TABLE 6.1. *(continued)*

Breed	Place of Origin	Date Introduced to U.S.	Performance Traits	Distinguishing Characteristics	Official Registration*
Ponies					
Connemara	Originated in Ireland, Probably from Iceland stock to which Andalusian, Barb, and Arab breeding was added.	1951	Ht.: 52–58 in. Wt.: 700–850 General purpose riding type. Some used as jumpers.	Most common horse colors exist. Piebalds and skewbalds ineligible for registry. Refined, chiseled head resembles Arabian.	American Connemara Pony Society, R.R. 2, Rochester, Ill. 62563. A permanent registry and a grading-up temporary registry are provided for.
Hackney	Developed as pony by parent horse association by introduction of Welsh pony breeding until studbook was closed in 1949.		Ht.: 11-2 to 14-2 Wt: varies with height. Strictly society show harness pony, comparable in every respect to Hackney horse except for size.	Same as Hackney horse except for size.	American Hackney Horse Society, 527 Madison Ave., Room 725, New York, N.Y. 10022. Closed studbook, and limited to maximum height of 14-2.
Pony of the Americas (Usually referred to as P.O.A.)	Developed in U.S., mainly by combining Appaloosa horse with Welsh and Shetland pony stock.		Ht.: 46-52 in. Conformation resembles small intermediate between Arabian and Quarter Horse. All-purpose western-type riding pony.	Appaloosa color including eyes, skin, and hoofs. Intermediate to large in size. Maximum height for permanent register 52 in.	Pony of the Americas Club, Inc, P.O. Box 1447, Mason City, Iowa 50401. Formed in 1954. Open studbook with both permanent and tentative registry provisions. Appaloosa color is essential.
Shetland	Originated in Shetland Isles north of Scotland. Used for draft in coal mines in British Isles.	1850	Ht.: 38–46 in. Wt.: 300–400 U.S. accepts higher maximum than Canada or England for registered show type. Selection in U.S. has been away from chunky "island type" children's pet toward refined, Hackney-type society show animal.	All common horse colors are prevalent. Solid colors. Refined, light harness type with alertness and flashy action.	American Shetland Pony Club, P. O. Box 2339, West Lafayette, Ind. 47906. Formed in 1888. Closed studbook for pure Shetlands. Club also maintains a registry for Harness Show Ponies resulting from crosses with other pony breeds.

TABLE 6.1. *(continued)*

Breed	Place of Origin	Date Introduced to U.S.	Performance Traits	Distinguishing Characteristics	Official Registration*
Welsh	Developed in Wales where it is known as Welsh Mountain Pony. Pure stock during most of its developmental period.	1884	Ht.: 56 in. maximum Actually a small horse, intermediate in height and weight between a light horse and Shetland pony. Conformation resembles a coach horse. Versatile for use as children's pet, showing, or pleasure riding and driving.	Usually gray, roan, black, brown, or chestnut. White or dilute colors, but no piebald or skewbald acceptable. Very hardy.	Welsh Pony Society of America, Inc., 202 N. Church St., West Chester, Pa. 19380. Formed in 1906 under different name. Closed studbook. Registers in groups "A" and "B" based on maximum heights of 50 and 56 in. respectively.
Draft Horses					
Belgian	Belgium	1886	Ht.: 15-2 to 17-0 Wt.: 1,900–2,200	Usually sorrel, chestnut, or roan. Heaviest muscled of U.S. draft breeds.	Belgian Draft Horse Corp. of America, Wabash, Ind. 46992. Organized 1887, under present name since 1937. Closed studbook.
Clydesdale	Clyde River Valley in South West Scotland	1842 to Canada, later to U.S.	Ht.: 16-0 to 17-0 Wt.: 1,700–2,000	Mostly bays or browns with white on face and legs. Long hair ("feather") on back of legs below knee and hock. Unusual style and action. Appear tall and leggy. A favorite in exhibition hitches.	Clydesdale Breeders' Ass. of the United States, Plymouth, Ind. 46563. Organized 1879. Closed studbook.
Percheron	France	1816 via Canada	Ht.: 16-0 to 17-0 Wt.: 1,700–2,000	Black with white markings at extremities or dappled gray most common. Intermediate in size and action to Belgian and Clydesdale.	Percheron Horse Ass. of America, Belmont, Ohio 43718. Organized in 1876, under different name. Closed studbook.
Shire	England	1836 to Canada, later to U.S.	Ht.: 16-0 to 17-2 Wt.: 1,800–2,300	Bay most common. Roman nose. "Feather" on legs like Clydesdale. Largest of draft breeds.	American Shire Horse Ass., Lynden, Wash. 98264. Organized 1875. Closed studbook.

TABLE 6.1. (continued)

Breed	Place of Origin	Date Introduced to U.S.	Performance Traits	Distinguishing Characteristics	Official Registration*
Suffolk (once called Suffolk Punch)	Suffolk County, England	1880	Ht.: 15-2 to 16-2 Wt.: 1,500–2,000	Always some shade of chestnut. Smallest of American draft breeds.	American Suffolk Horse Ass., Lynden, Wash. 98264. Organized in 1911. Chestnut color a registry requirement. Closed studbook.
Asses American Jack	Principally Spain	1787	Ht.: 15-0 to 16-0 Wt.: 1,050–1,200	Black with white hair around eyes, nose, and underline. Some solid reds. Gray disliked. Long, large ears and Roman nose. Mane and tail hair thin or absent.	Standard Jack and Jennet Registry of America, Route 7, Lexington, Ky. 40502. Bred entirely for use in producing mules. Variable closed or open studbook requirements have obtained.
Miniature Donkey	Mediterranean areas		Ht.: 38 in. or less Pack animal typical of those widely used in South America. Used in U.S. as children's pet or hobby animal.	Variable color, usually gray or reddish brown to almost black. Darker dorsal stripe and stripe over withers. Typical ass ears, head profile, mane, and tail.	Miniature Donkey Registry, 1108 Jackson St., Omaha, Neb. 68102. Established in 1958. Closed studbook, with 38 in. height maximum and specified color requirements.

ARABIAN

Half-Arab and Anglo-Arab Registries
224 East Olive Ave.
Burbank, Calif. 91503

International Arabian Horse Ass.
224 East Olive Ave.
Burbank, Calif. 91503

PONY

American Quarter Pony Ass.
New Sharon, Iowa 50207

American Walking Pony Ass.
Route 5, Box 64–A
Upper River Road
Macon, Ga. 31201

Cross-Bred Pony Registry
1108 Jackson St.
Omaha, Neb. 68102

National Appaloosa Pony, Inc.
112 East Eighth St.
P.O. Box 297
Rochester, Ind. 46975

National Trotting Pony Ass., Inc.
575 Broadway
Hanover, Pa. 17331

Pinto Pony Registry
P.O. Box 1373
Visalia, Calif. 93277

Shetland Pony Identification Bureau, Inc.
1108 Jackson St.
Omaha, Neb. 68102

U.S. Trotting Pony Ass.
P.O. Box 1250
Lafayette, Ind. 47902

QUARTER HORSE

Model Quarter Horse Ass.
P.O. Box 396
Lincoln, Calif. 95618

National Quarter Horse Registry, Inc.
Raywood, Tex. 77582

Original Half Quarter Horse Registry
Hubbard, Ore. 97032

Standard Quarter Horse Ass.
4390 Fenton St.
Denver, Colo. 80212

THOROUGHBRED

American Thoroughbred Breeders &
Owners Ass.
1736 Alexandria Dr.
P.O. Box 4038
Lexington, Ky. 40504

Thoroughbred Racing Ass. of U.S., Inc.
Suite 919
220 East 42nd St.
New York, N.Y. 10017

GENERAL HORSES AND MULES

American Andalusian Ass.
P.O. Box 1290
Silver City, N.Mex. 88061

American Ass. of Owners & Breeders of
Peruvian Paso Horses
P.O. Box 371
Calabasas, Calif. 91302

American Buckskin Registry Ass., Inc.
P.O. Box 1125
Anderson, Calif. 96007

American Donkey & Mule Society
Rt. 1, Box 519–A
Denton, Tex. 76201

American Gotland Horse Ass.
R.R. #2, Box 181
Elkland, Mo. 65644

American Indian Horse Registry, Inc.
P.O. Box 9192
Phoenix, Ariz. 85020

American Mustang Ass.
P.O. Box 9243
Phoenix, Ariz. 85020

American Part-Blooded Horse Registry
4120 S.E. River Dr.
Portland, Ore. 97222

American Paso Fino Pleasure Horse Ass., Inc.
Arrott Building
401 Wood St.
Pittsburgh, Pa. 15222

International Buckskin Horse Registry
P.O. Box 772
Anderson, Calif. 96007

Missouri Fox Trotting Horse Ass.
Ava, Mo. 65608

National Ass. of Paso Fino Horses of
Puerto Rico
Aptdo. 253
Guaynabo, P.R. 00657

National Cutting Horse Ass.
P.O. Box 12155
Fort Worth, Tex. 76116

National Mustang Ass.
Newcastle, Utah 84756

Peruvian Paso Horse Registry of
North America
12670 Skyline Blvd.
Woodside, Calif. 94062

Spanish Mustang Registry, Inc.
1065 East Lehi Road
Mesa, Ariz. 85203

Trakehner Breed Ass. & Registry of
America, Inc.
Rt. 1, Box 177
Petersburg, Va. 23802

DRAFT BREEDS

Cleveland Bay Society of America
White Post, Va. 22663

FIG. 6.6. (left) *Arabian mare, Dornaba, National Champion and Legion of Merit winner, owned by Dr. Howard F. Kale, Wash.* (right) *Arabian stallion, Bask, National Champion and Legion of Merit winner, owned by Lasma Arabians, Ariz.* (International Arabian Horse Assn.)

FIG. 6.7. *Thoroughbred stallion, Buckpasser, by Tom Fool. Horse of the Year and Best 3-year-old Handicap Horse in 1966.* (New York Racing Assn., Inc.)

FIG. 6.8. *Champion American Saddle Horse mare, Denmark's Daydream.* (American Saddle Horse Breeders Assn., Inc.)

FIG. 6.9. *American Albino stallion, Snow King. 1970 National Champion Stallion.* (American Albino Assn., Inc.)

FIG. 6.10. *Appaloosa champion stallion, Captain Barry.* (Appaloosa Horse Club)

FIG. 6.11. *Morgan champion mare.* (Morgan Horse Club, Inc.)

FIG. 6.12. (left) *Paint mare, Bandit's Squaw, APHA Champion. Overo color pattern.* (right) *Paint stallion, Joe Chief Bar, APHA Supreme Champion. Tobiano color pattern.* (American Paint Horse Assn.)

FIG. *6.13. Pinto gelding with tobiano color pattern.* (Pinto Horse Assn. of America, Inc.)

FIG. *6.14. Idealized color pattern of Palomino.*

FIG. *6.15. Quarter Horse mare, Poco Lena.*

FIG. *6.16. Tennessee Walking Horse stallion, Shadow's Royal Flush, 3-year-old World Champion.* (Courtesy H. C. Bailey, Jackson, Miss., owner along with Buford Chitwood, Calhoun, Ga., and Mack Motes, Eagleville, Tenn.)

FIG. *6.17. Standardbred stallion, Gene Abbe, famous sire of money winners.* (United States Trotting Assn.)

FIG. 6.18. *Connemara pony stallion, Marconi.* (American Connemara Pony Society)

FIG. 6.19. *Hackney pony mare and foal.* (American Hackney Horse Society)

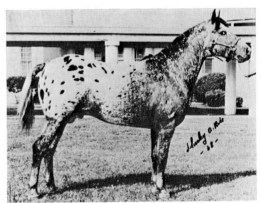

FIG. 6.20. (left) *POA mare.* (right) *POA stallion.* (*Note variations of the Appaloosa-type color pattern.*) (Pony of the Americas Club, Inc.)

FIG. *6.21. Shetland pony stallion.* (American Shetland Pony Club)

FIG. *6.22. Welsh pony mare, Liseter Starlight.* (Liseter Hall Farm, Pa.)

FIG. *6.23. Belgians showing at a state fair.*

FIG. *6.24. Clydesdale mare.*

FIG. *6.25. Percheron stallion.*

FIG. *6.26. Shire stallion.*

FIG. 6.27. *Suffolk mare.*

FIG. 6.28. (left) *American Jack.* (right) *Mules sired by American Jacks.* (Willis Grumbein, Kan.)

FIG. 6.29. *Registered Miniature Donkey—a unique Christmas present.*

READING LIST

American Shetland Pony Journal
    P.O. Box 1250, Lafayette, Ind. 47902
Appaloosa News
    P.O. Box 403, Moscow, Idaho 83843
Arabian Horse News
    P.O. Box 1009, Boulder, Colo. 80302
Arabian Horse World
    23 E. Main St., Springville, N.Y. 14141
Arizona Horseman
    2517 N. Central, Phoenix, Ariz. 85004
Belgian Review
    282 S. Wabash St., Wabash, Ind. 46992
Bit and Bridle
    122—40th St., Toronto 14, Canada
Blood Horse
    P.O. Box 4038, Lexington, Ky. 40504
Canadian Horse
    P.O. Box 127, Rexdale, Ontario, Canada
Chronicle of the Horse
    Middleburg, Va. 22117
Draft Horse Journal
    1803 Oakland Ave., Cedar Falls, Iowa 50613
Harness Horse
    P.O. Box 1831, Harrisburg, Pa. 17105
Horse and Show Journal
    20614 Aurora Road, Bedford, Ohio 44146
Horse World
    P.O. Box 588, Lexington, Ky. 40501
Horseman
    5314 Bingle Road, Houston, Tex. 77018
Horseman's Advisor
    624 Payton Ave., Des Moines, Iowa 50315
Horseman's Journal
    425 13th St. N.W., Washington, D.C. 20004
International Trotter and Pacer
    5221 E. Simpson Ferry Rd., Mechanicsburg, Pa. 17055
Morgan Horse
    Leominister, Mass. 01453
Mustanger Stirr Ups
    P.O. Box 9243, Phoenix, Ariz. 85020
Paint Horse Journal
    P.O. Box 12487, Fort Worth, Tex. 76116
Palomino Horses
    P.O. Box 249, Mineral Wells, Tex. 76067
Pinto Horse
    4315 Hilldale Road, San Diego, Calif. 92116
Pony of the Americas
    P.O. Box 1447, Mason City, Iowa 50401
Quarter Horse Journal
    P.O. Box 9105, Amarillo, Tex. 79105
Quarter Racing Record
    Box 2473, Fort Worth, Tex. 76101

Rodeo News
    P.O. Box 587, Pauls Valley, Okla. 73075
Saddle and Bridle
    8011 Clayton Road, St. Louis, Mo. 63117
Thoroughbred Record
    Box 580, Lexington, Ky. 40501
Voice of the Tennessee Walking Horse
    P.O. Box 6009, Chattanooga, Tenn. 37401
Welsh News
    1770 Lancaster Ave., Paoli, Pa. 19301
Western Horseman
    3850 N. Nevada Ave., Colorado Springs, Colo. 80901
Your Pony and Hackney Journal
    1040 W. James St., Columbus, Wis. 53925

[CHAPTER SEVEN] Poultry

[GALLINACEOUS AND ANSERINE BIRDS]

IT IS ESTIMATED that there is a combined total of more than 3 billion chickens, turkeys, ducks, and geese in the world. The production of poultry meat and chicken eggs accounts for nearly 20% of the cash farm income in the United States; poultry meat, along with beef, has recorded the largest increases in per capita consumption among food products produced from animals. Chicken broiler and egg production rank second only to commercial milk production in efficiency of converting feed to human food nutrients—well ahead of other livestock enterprises.

The ancestor of the chicken (*Gallus gallus* or *Gallus domesticus*) is the wild jungle fowl *Gallus bankiva,* found in the jungles of India. The two clutches totaling about 30 eggs per year laid by these fowl is a far cry from today's commercial laying hens that average some 220 eggs, with individuals reaching well above 300 eggs per year. Chickens are widely distributed throughout the world and display an immense variety of body types, plumage patterns, head furnishings, and other characteristics. The breeding and management methods employed with chickens run the gamut from the small family flock associated with subsistence agriculture to the most highly specialized and automated production "factory" consisting of many thousand birds.

The turkey, Muscovy duck, and Canada goose are natives of the Western Hemisphere. Domesticated turkeys are descendants of the North American wild turkey *Meleagris gallopavo,* classified in the Meleagrididae family within the order Galliformes. Several geographic regional subspecies which interbreed freely are described in this family. These wild turkeys formerly ranged over most of Mexico, the United States, and southern Ontario, Canada. The ocellated turkey of the Yucatan Peninsula of Mexico is a beautiful plumaged bird of a different genus, *Agriocharis ocellata,* and does not enter into the ancestry of the common turkey. Domesticated turkeys are raised for their meat.

All domesticated ducks except the Muscovy are descended from the wild mallard *Anas platyrhynchos,* of the family Anatidae in the order Anseriformes. The Muscovy is *Cairina moschata* of the same family. Like chickens, domesticated ducks have been developed along three main lines: (1) those bred for meat, (2) those bred for egg production, and (3) those bred for ornamental or exhibition purposes. While the best egg-laying breeds of ducks usually excel the best chicken layers in egg production, ducks are produced principally for meat in the United States because of the limited market for duck eggs. Duck meat and eggs are more popular in other countries than the United States, and much of the demand for duck meat in the United States is in the larger cities among people of more recent foreign extraction. Two small importations of Pekin ducks were brought from China in 1873 and provided the foundation for nearly all commercial duck meat production in the United States, replacing other breeds brought from Europe by the early settlers. The Muscovy, a different species, originated in Brazil and was introduced into the United States between 1840 and 1850. The incubation period for Muscovy eggs is 33–35 days compared to 28 days for the other domesticated breeds of ducks. The Muscovy also differs from the other breeds in exhibiting a marked size difference between male and female, the male being about one-third larger than the female.

The domestic goose *Anser anser* belongs to the subfamily Anserinae, family Anatidae, order Anseriformes. The goose has been revered and cherished as a family possession since ancient times. Geese may live for 30 years; they become very possessive and guard their premises in much the same way as a watchdog. Geese are not very important commercially as either meat or egg producers in the United States. Their propensity for grazing on weeds has made them commercially useful in ridding strawberry fields, cotton fields, or other cultivated crops of weeds; their down feathers have always been prized for making pillows. Domesticated breeds of geese were introduced into the United States by the early settlers, having originated in Europe and Asia.

The guinea fowl *Numida meleagris*, of the family Numididae in the order Galliformes, originated in Africa. Small flocks of these noisy "watchdogs" were frequently found along with other poultry on early American farmsteads. They have not become commercially important, although their "gamey" type of meat is considered a delicacy by some people. Their temperament probably has been one drawback to their adaptation to intensive commercial production methods.

Mention should also be made of pheasants, quail, and other game birds that are produced under confinement to be released to the wild for hunting. The recent introduction of the Japanese quail *Coturnix coturnix* deserves mention as one of the most useful laboratory research animals. Its small size, adaptability to confinement, and very rapid reproductive rate (females begin laying at 6–8 weeks of age) are important factors in this utility.

USE OF BREEDS FOR COMMERCIAL PRODUCTION

The present-day breeding methods used in production of commercial chickens represent the furthest step along the path of genetic technology of any domesticated species. The breeding of other improved live-stock tends to be moving along the same route to varying degrees, so a brief chronology of the steps taken may provide perspective for the future.

1. A great proliferation of breeds and varieties within breeds took place, with selection based on readily observable phenotypic character-istics such as body form, plumage color and patterns, and type of comb.

2. Some measures of performance such as egg production, growth rate, or meatiness entered into the selection within a breed and variety.

3. Crossing of lines selected on the basis of performance or pheno-type was initiated.

4. The combining ability of breed crosses was evaluated on the basis of performance of the crossbred progeny.

5. The crossing or "hybridization" concept was further refined by the development of inbred lines and the crossing of those whose progeny performed best.

6. Reciprocal and reciprocal recurrent matings came into use on the basis of performance-tested crossline or crossbred progeny.

Since the employment of these advanced techniques requires cen-tralized control of breeding and performance evaluation of seed stock, under the guidance of well-trained geneticists, a hierarchial breeding structure is well developed in the production of commercial chickens. The following sketch illustrates this structure, in which can be rec-ognized emerging parallels with some current swine-breeding organiza-tions and to a less advanced degree with cattle-breeding efforts.

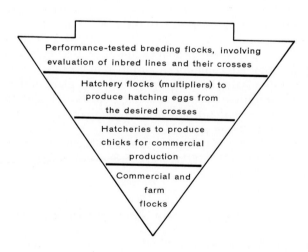

A significant development in early efforts to more fully utilize the available genetic resources and knowledge was the initiation in 1935 of the National Poultry Improvement Plan. This program is a joint effort between official state agencies and the USDA. A comparable program for turkeys is the National Turkey Improvement Plan. These improvement programs provide for official recognition and standardization of seed stock sources based on performance testing.

The trap nest was invented in 1895 by James E. Rice, a professor at Cornell University, and its promotion resulted in many individual egg-laying performance records. This methods was subsequently replaced by centralized random sample performance tests involving stocks produced at the test station from eggs supplied by several different breeders. This latter concept is not unlike the Danish progeny testing stations for dairy cattle and the central swine testing and beef bull testing stations. Similar testing methods are widely employed for broiler stock. The extent of improvement in egg production that has occurred is evident from USDA crop reporting statistics, which showed 209 eggs produced per layer on hand during 1960 compared to 112 eggs in 1925. Improvements in disease control and other environmental influences of course share credit with genetic improvement for this gain.

For the production of poultry meat—chicken, turkey, or duck—a strong preference exists for white-feathered birds due to their freedom from dark pinfeathers and consequent greater ease in producing attractive carcasses. Color inheritance in poultry offers interesting genetic problems, since some chicken breeds are "dominant whites," others "recessive whites," and some color genes are sex-linked. Most U.S. markets also prefer white-shelled eggs. These preferences, along with performance characteristics, help to explain the heavy emphasis on the White Wyandotte and White Plymouth Rock as bases for broiler stock, the White Holland and Beltsville White for turkey production, the White Pekin duck, and the White Leghorn background of commercial chicken egg-laying stock. Infusions of other breeds have been used for specific traits, such as the Cornish to improve carcass conformation and quality in broiler stock.

It seems obvious that the available choices of well-established breeds and varieties are infinitely greater than the number that are currently influential in commercial production. This is true even if the choice were limited to those breeds and varieties developed along utilitarian lines, to say nothing of the novelty types such as the bantams, the long-tailed exhibition chickens of Japan, and the fighting chickens popular and legal for cockfights in many Spanish-speaking countries. To what extent the evolution of seed stock production for commercial use in other species of livestock will ultimately parallel that of the chicken industry is a matter of conjecture, but present trends certainly point in that direction.

STANDARDIZATION OF BREEDS AND VARIETIES

The poultry fanciers have hardly been rivaled by any other group in the precision of setting up official judging score cards, depicting ideal types pictorially, detailing disqualifying defects, and other activities associated with fancy points of breed perfection and exhibition.

The American Poultry Association, representing United States and Canadian breeders, was organized at Buffalo, New York, in February 1873 with the primary objective of standardizing the varieties of domestic poultry. The following principles were adopted in this effort:

1. In each breed the most useful type should be made the Standard type.

2. Recognized breeds should be easily identifiable by at least one conspicuous character or combination of same.

3. Recognition of color varieties within a breed should be limited to plainly distinctive color patterns.

The first *American Standard of Excellence,* later titled *American Standard of Perfection,* was issued in February 1874. This consisted of verbal specifications until 1905, when the practice of supplementing the descriptions with professional drawings or color paintings was first introduced.

An abbreviated summary of recognized breeds and varieties is given in Table 7.1.

TABLE 7.1. Summary of Poultry Breeds

Class and General Characteristics	Breed	Place of Origin	Specific Breed Information
Chickens (12 classes. Many breeds have subvarieties [based mainly on color and comb type] and bantam types. Some of the most recently introduced or recognized breeds or varieties are not listed here.)			
Class 1. American	Plymouth Rock	U.S.	7 color varieties, 2 bantam types. White Plymouth Rock important in modern commercial broiler strains.
Meat or dual-purpose type; yellow skin; all except Lamona lay brown-shelled eggs of varying shades.	Wyandotte	U.S.	8 color varieties, 6 bantam types. Rose comb.
	Java	Imported from Isle of Java in 1883.	Black and mottled color varieties.
	Dominique	U.S.	Dark and light barred plumage. Rose comb.
	Rhode Island Red	U.S.	Single and rose comb varieties, 1 bantam type.
	Rhode Island White	U.S.	Rose comb.
	Chantecler	Quebec, Canada	White and partridge color varieties. Small, cushionlike comb.
	Jersey Black Giant	U.S.	Also a white variety of more recent origin.
	Lamona	USDA, Beltsville	White-shelled eggs.
	New Hampshire	U.S. (from R.I. Red stock)	Red plumage.
Class 2. Asiatic	Brahma	China	3 color varieties, 2 bantam types.
Heavy meat or exhibition type; yellow skin except for Langshan; feathered legs and toes to varying degrees; lay brown-shelled eggs.	Cochin	China	4 color varieties, 4 bantam types.
	Langshan	China	White skin. 2 color varieties.
Class 3. Mediterranean	Leghorn	Italy	White variety most influential foundation breed in modern commercial egg-laying strains. 12 varieties based on color and single or rose comb, 1 bantam type.
Lighter weight, egg-laying type; classed as nonsitters (nonbroody); lay white-shelled eggs. Skin color variable among breeds and varieties.	Minorca	Spain	5 varieties based on color and single or rose comb, 1 bantam type.

Table 7.1. *(continued)*

Class and General Characteristics	Breed	Place of Origin	Specific Breed Information
	Spanish	Spain	Oldest Mediterranean breed. Also called White-Faced Black Spanish to distinguish from Minorca, which formerly was called Red-Faced Black Spanish.
	Blue Andalusian	Spain (Andalusia Province)	Heterozygous color pattern inheritance.
	Ancona	Italy	Single and rose comb varieties.
	Buttercup	Sicily	....
Class 4. English Includes meat, dual-purpose, and egg-laying breeds.	Dorking	Rome, Italy	3 color varieties. Dual-purpose. One of most ancient breeds, brought to Great Britain by Romans at time of Julius Caesar, and described well before that time. White-shelled eggs.
	Redcap	Derbyshire, England	Dual-purpose. White-shelled eggs. Uniquely large rose comb.
	Orpington	Orpington, England	4 color varieties. Dual-purpose. Light to dark brown-shelled eggs.
	Cornish	Cornwall, England	Heavy meat type, with **same body conformation in male and female**. Important genetic ingredient in some modern commercial broiler chickens. 4 color varieties, 3 bantam types. Brown-shelled eggs.
	Sussex	Sussex County, England	3 color varieties. Dual-purpose. Brown-shelled eggs.
	Australorp	Australia	Developed from Orpingtons, principally for egg production. **Eggshells slightly brown-tinted.**
Class 5. Polish Crested, nonsitter, ornamental type.	Polish	Eastern Europe	9 varieties based on color, crest, and beard—each with a bantam type. Strictly an exhibition breed. White-shelled eggs.
Class 6. Hamburgs Nonsitter, ornamental type.	Hamburg	Netherlands via Hamburg, Germany	6 color varieties, 1 bantam type. White-shelled eggs.
Class 7. French General purpose; white skin; all except Faverolles produce white-	Houdan	Houdan, France	Crested. 2 color varieties. Also known as Normandy fowl.
	Crevecoeurs	France	Crested. Large size. Probably a sub-

TABLE 7.1. *(continued)*

Class and General Characteristics	Breed	Place of Origin	Specific Breed Information
	LaFleche	LaFleche, France	Has trace of crest. Black.
	Faverolles	France	Lay light brown-shelled eggs. Developed from crossing several older breeds. Salmon color.
Class 8. Continental Nonsitter, egg-laying type; lay white-shelled eggs.	Campine	Belgium	2 color varieties.
	Lakenvelder	Germany	Old established breed. Black & white color.
Class 9. Games High-stationed, stylish carriage. Used mostly for exhibition rather than pit fighting for which they were originally developed.	Modern Game	England	8 color varieties, 8 bantam types.
	Old English Game	England	8 color varieties, 6 bantam types.
Class 10. Orientals Variable types and uses.	Sumatra	Sumatra	Black. Yellow skin. Pea comb. Dual-purpose and exhibition breed.
	Malay	Asia Brought to England as early as 1830.	Large, tall breed with distinctive wide skull. Meat and exhibition type. 1 bantam type.
	Cubalaya	Cuba	Hardy, meat type. 3 color varieties.
Class 11. Ornamental Bantams Bred strictly for exhibition. Clean-legged and feather-legged types.	*Clean-legged* Sebright	England	2 color varieties. Rose comb.
	Rose-Comb	Europe	2 color varieties.
	Japanese	Japan	4 color varieties.
	Feather-legged Mille Fleur	Europe	Bearded and nonbearded types.
	Silkie	China & Japan	Unique feather structure. Bearded and nonbearded types.
	Booted	Probably Asia	White. Dark blue skin.
Class 12. Miscellaneous Novelty, exhibition breeds.	Sultan	Developed in Turkey from southeastern European origin.	Crested and bearded.
	Frizzle	Probably India	Unique feather structure, showing backward curling.

TABLE 7.1. *(continued)*

Ducks (12 breeds including meat, egg, and ornamental types)

Class 13.

Class and General Characteristics	Breed	Place of Origin	Specific Breed Information
	Pekin	China	White. Yellow skin. Meat type. Account for virtually all commercial U.S. duck meat production.
	Aylesbury	England	Very similar to Pekin.
	Rouen	France	Eggs blue-shelled or tending toward white. Somewhat like Mallard with difference in plumage colors between male and female. Meat type.
	Cayuga	U.S.	Greenish-black plumage. Eggs black-shelled at start of laying period, later becoming blue. Medium size.
	Call	....	Bantam-sized. Gray and white varieties, miniatures of Rouen and Pekin respectively.
	East India	Europe (probably Netherlands & Belgium)	Bantam-sized. Greenish-black plumage. Black-shelled eggs at start of production.
	Muscovy	Brazil	Two varieties, colored and white. Different species than other breeds. Large size, with male larger than female. 35-day incubation period compared to 28 days for other ducks.
	Swedish	Sweden	Medium size. Blue plumage.
	Buff	....	Medium size. Buff or seal-brown plumage.
	Crested	....	White, crested, small size.
	Runner (or Indian Runner)	Europe (probably Netherland & Belgium)	Small. Upright stance. Very high producer of white-shelled eggs. 3 color varieties
	Khaki Campbell	Netherlands	Small. Khaki-colored plumage, differing between male and female. Very high egg producers.

TABLE 7.1. *(continued)*

Class and General Characteristics	Breed	Place of Origin	Specific Breed Information
Geese (8 breeds including utilitarian and novelty types. Weights listed refer to adult female and male respectively.)			
Class 14. All except Canada originated in Eastern Hemisphere.	Toulouse	Europe	Heavy weight (20, 26 lb.). Gray color shades.
	Emden	Europe	18, 20 lb. White.
	African	Europe	18, 20 lb. Brown color shades, black bill with knob at junction with head.
	Chinese	Europe	Small size (10, 12 lb.). Brown or white. Knob like African.
	Canada	Domesticated from wild Canada geese.	Small size (10, 12 lb.). Mainly ornamental.
	Egyptian	Europe	Very small (4½, 5½ lb.). Ornamental. Pugnacious temperament.
	Sebastopol	Turkey	One of oldest domesticated varieties. 12, 14 lb. Curled feathers.
	Pilgrim	From stock brought to America by Pilgrims in 1620.	13, 14 lb. White with specified gray areas. Meat producer.
	American Buff	U.S.	16, 18 lb. Buff color shades.
Turkeys (Native of North America. Raised for meat. Eggshells typically brown speckled in all breeds. Six varieties. Weights listed refer to adult female and male respectively.)			
Class 15.	Bronze	....	Color pattern similar to wild progenitor. Old domesticated variety. 20, 36 lb.
	Narragansett	....	Darker than Bronze. 18, 33 lb.
	White Holland	....	White plumage. Basic variety for most popular commercial strains. 18, 33 lb.
	Black	....	Black, metallic, lustrous plumage. 18, 33 lb.
	Slate	....	Slaty or ashy blue color. 18, 33 lb.
	Bourbon Red	....	Distinctive red shades (mahogany) with white wing tips and tail. 18, 33 lb.

FIG. 7.1. *Jungle Fowl* Gallus bankiva. (Painting of actual. Courtesy *Poultry Tribune*)

FIG. 7.2. *Commercial egg layers, white-shelled eggs.* (Babcock Poultry Farm, Inc.)

FIG. 7.3. *Commercial egg layer, brown-shelled eggs.* (Babcock Poultry Farm, Inc.)

FIG. 7.4. *Commercial broiler parent stock.* (Welp Breeding Farm)

FIG. 7.5. *Sex-linked color pattern. Crossing New Hampshire male with Barred Plymouth Rock female produces barred male progeny and solid reddish-colored female progeny.* (Painting of actual. Courtesy *Poultry Tribune*)

FIG. 7.6. *American breeds and varieties.*

(A) White Plymouth Rocks

(B) Partridge Plymouth Rocks

(C) Silver-Penciled Plymouth Rocks

(D) White Wyandottes

Note: All poultry illustrations are paintings of actual types and are courtesy of Poultry Tribune.

(E) Columbian Wyandottes

(F) Single Comb Rhode
Island Reds

(G) Jersey Black Giants

(H) Lamonas

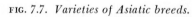

FIG. 7.7. Varieties of Asiatic breeds.

(A) Light Brahmas

(B) Buff Cochins

(C) White Langshans

FIG. 7.8. Mediterranean breeds and varieties.

(A) Single Comb White Leghorns

(B) Single Comb Brown Leghorns

(C) Single Comb Black Minorcas

(D) Single Comb Blue Andalusians

(E) Single Comb Anconas   (F) Buttercups

FIG. 7.9. English breeds and varieties.

(A) Silver Gray Dorkings   (B) Buff Orpingtons

(C) Dark Cornish

(D) White Laced Red
Cornish

(E) Speckled Sussex

(F) Australorps

FIG. 7.10. White Crested
Black Polish.

FIG. 7.11. Silver Spangled
Hamburgs.

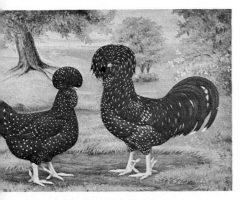

FIG. 7.12. Mottled Houdans.

FIG. 7.13. Silver Campines.

FIG. 7.14. Old English Games (left to right): Golden Duckwing female and Black Breasted Red male.

FIG. 7.15. Ornamental Bantams.

(A) Golden Sebrights

(B) Mille Fleur Booted

FIG. 7.16. Breeds of Ducks.

(A) Pekins

(B) Rouens

(C) Colored Muscovys

(D) Runners (Fawn & White and White varieties)

FIG. 7.17. *Breeds of Geese.*

(A) (left) *Emden,* (right) *Toulouse*

(B) African

FIG. 7.18. *Breeds of Turkeys.*

(A) Bronze

(B) Narragansett

148

(C) White Holland

(D) Bourbon Red

READING LIST

Jull, Morley A. 1936. Superior breeding stock in poultry. *USDA Yearbook of Agriculture, 1936.* U.S. Government Printing Office, Washington, D.C.

Lee, A. R. 1937. Duck breeding. *USDA Yearbook of Agriculture, 1937.* U.S. Government Printing Office, Washington, D.C.

Marsden, S. J., and C. W. Knox. 1937. The breeding of turkeys. *USDA Yearbook of Agriculture, 1937.* U.S. Government Printing Office, Washington, D.C.

[CHAPTER EIGHT] Arctic Livestock

T HE NORTHWARD PIONEERING of the United States and Canada into the arctic regions has been accompanied by an increasing awareness of the native citizens of the Far North. As these indigenous populations shift from a hunting and fishing economy, the husbandry of livestock is a logical step toward providing a more stable subsistence. The domesticated reindeer had long been an important animal in the arctic economy of Lapland and Siberia prior to its introduction to Alaska. More recently the possible utility of the musk ox has stimulated interest in its domestication.

REINDEER

When the Territory of Alaska became the 49th state in 1959, the reindeer *(Rangifer tarandus)* became a definite part of the enormous animal industry of the United States. The domestic reindeer of Alaska descended from the Lapland reindeer and is handled and cared for like other range animals. It should not be confused with wild reindeer or caribou which also exist in Alaska. According to the USDA there were 45,000 head of domestic reindeer in Alaska in 1962. The numbers of other farm animals in Alaska on this same date were: 7,900 cattle and calves, 15,000 sheep, and 1,000 hogs. During 1961 Alaska produced 309,000 lb. beef, 135,000 lb. pork, and 485,000 lb. domestic reindeer meat.

The reindeer of Alaska, like all other important farm animals in North America, are not native to the continent. They were imported into Alaska to improve the economy of the far northern area. The first importation was made in 1891 and consisted of only 18 head. They were brought in from Lapland by Dr. Sheldon Jackson, an early missionary and educator in Alaska. In 1892 the United States government imported 172 animals. By 1902 a total of 4,795 reindeer had been imported from Lapland and Siberia for use by the natives as food, clothing, and transportation.

Domestic reindeer have thrived in Alaska. They relish the grasses, seaweeds, and rock lichen and can scrape away several feet of snow with

FIG. *8.1. Mature male Laplander reindeer.* (V. B. Scheffer, U.S. Fish & Wildlife Service)

their horns and hoofs to locate feed if necessary. The USDA has stated that there were 200,000 sq. mi. of rangeland in Alaska well suited to raising reindeer. L. J. Palmer in 1934 estimated the carrying capacity of this area was at least 4 million head. Recent experiences in reindeer production cause many to consider this figure far too high. From 1902 to 1930 reindeer numbers in Alaska increased from a few thousand to 641,000. This number remained constant until 1936 and then declined rapidly until 1950 when only 25,000 reindeer remained. The reasons for the population decline are not fully known; but overgrazing, wolves, improper herding, and marketing appear to have been major contributing factors. Since 1950 numbers have constantly increased as better care and management have been provided. However, H. C. Hanson, who studied the problem in 1952, thinks the carrying capacity of the good reindeer range is only about 320,000 head.

Two species of wild reindeer or caribou are native to the northern part of North America, living in both woodland and barren country. Wild reindeer will occasionally join the herds of domestic reindeer and will cross with them. Producers do not like the crosses or the intermingling of wild and domestic reindeer; herds with crossbreds or wild reindeer among the domestic animals are hard to handle.

The domestic reindeer has not been specialized as have most other domestic animals. On the basis of use, therefore, the reindeer is a general-purpose animal with meat, skins, milk, and draft being of major importance. Today meat and skins are by far its greatest value. However, natives still use some reindeer for milk production, some as pack animals, and some for drawing sleds. Under good conditions a reindeer can draw a load of 300 lb. and travel at a rate of 100 mi. per day. In deep snow (18–24 in.) two reindeer will pull as heavy a load as will 10 dogs. A reindeer must eat frequently when used heavily, but his feed is at hand when traveling on reindeer range. All domestic reindeer in Alaska are owned by Alaskan natives or the United States federal government, according to a federal law adopted to aid the natives in making a living. It seems likely, however, that this law is greatly limiting the possibilities of developing a large, profitable reindeer industry.

IMPROVEMENT EFFORTS. Breeders have not developed specific breeds of reindeer, but they have selected some males for superior ruggedness. The domestic reindeer possess definite production characteristics which make them much more valuable than wild reindeer, but there is still need for improvement. An appropriate name for the reindeer now in Alaska would seem to be the Laplander, since the stock all came originally from Lapland.

CHARACTERISTICS. Mature animals weigh 300–500 lb. and stand 40–50 in. high at the withers. Fawns weigh 7–16 lb. at birth and grow rapidly, nourished by milk of about 34% solids content. The fat content of reindeer milk averages about 19%, compared to 3.5–5.5% for most cattle, sheep, and goats. Protein and mineral content are about 3 and 2 times as high, respectively, and lactose one-half as high as in average cow's milk. Reindeer have small udders with four teats. The gestation period is about 8 months.

Branching antlers, which are shed annually, are present in both sexes. The body is long and rather shallow, with a heavy neck and a sloping rump ending in a short, goatlike tail usually fringed with white hair. The legs are long and light boned; hoofs are solid black, large, and broad. The coat color is brownish, becoming very dark in summer and lighter in winter, with white markings common on hoof heads and underline.

There are no official breeds of reindeer nor associations for promoting them. Details on reindeer production and husbandry may be obtained from the USDA, P.O. Box 1737, Palmer, Alaska 99645; U.S. Dept. of Interior, Bureau of Indian Affairs, Juneau, Alaska 99801; or State of Alaska, Division of Agriculture, Palmer, Alaska 99645.

MUSK OXEN

Considerable mystery and conflicting information surround this shaggy denizen of the Far North. Both "musk ox" and its scientific name, *Ovibos moschatus,* coined in 1816 by French scientist De Blainville, appear to be misnomers. The musk ox has neither musk glands nor a musky odor, except for a characteristic sweetish odor arising from urine of the bulls during the rutting season. The generic name ovibos suggests something between cattle and sheep. Chromosome studies by William J. Tietz, Jr. of Colorado State University and John J. Teal, Jr. of University of Alaska, reported in 1967, show the 2n chromosome number to be 48, the same as the Indian water buffalo. European and zebu cattle, the European bison, the domestic goat, and the yak *(Poephagus grunniens)* all have a 2n chromosome number of 60. Earlier serological studies had indicated the musk ox to be more similar to sheep and goats than to cattle and bison. Reported courtship and other behavior patterns also suggest this similarity. No interspecies hybridization efforts with the musk ox have been reported.

Canadian Henry Kelsey reported the discovery of the musk ox in 1689. A prominent observation at that time was the gossamer, cashmere-like sheets of wool fiber shed by the animals in the spring, suggesting a possible commercial value. This undercoat, called *qiviut* by the Eskimos, is shed out between the outer guard hairs or can be easily plucked in a manner similar to that of camels. The unusual and valuable characteristics of quiviut have been confirmed by many studies since that time. The long, fine fibers accept all dyes readily and produce fabrics that are uniquely warm and light weight.

Musk oxen were at one time numerous in arctic regions. As with many animal species confronted by man, the musk ox fell prey to the exploitation of overzealous hunters until protective steps were taken to slow this onslaught. Their habit of forming a ring fringed by the herd bulls was a more effective defense against wolves and other natural enemies than against the weaponry of man. Several early observers suggested domestication of the species. The most devoted advocate was arctic investigator Vilhjalmur Stefansson, progenitor of the domestication project that ultimately came into being in 1954 under the farsighted direction of John J. Teal, Jr., with the establishment of a small herd of wild musk oxen on his farm in Vermont. In 1964, as director of the Institute of Northern Agriculture Research (INAR), Huntington Center, Vt., Teal established the first musk ox breeding station at the University of Alaska, with support from the W. K. Kellogg Foundation. Ten additional female calves were captured on Nunivak Island, where earlier introductions had been made from Greenland in the 1930s, and were added to the University herd in 1965.

Three musk ox domestication stations have been established by INAR: in Alaska, northern Quebec, and Norway.

COMMERCIAL OBJECTIVES. The principal objective of musk ox domestication was to establish a village-based industry for remote Eskimo groups, based on quiviut production. The idea was to develop breeding stocks and husbandry practices for village ownership of herds of the *oomingmak* (the bearded one), as the musk ox is called by the Eskimos, and to use the talents of village women to produce scarves, shawls, or other garments from the quiviut. Investigation of the fiber by textile

FIG. 8.2. *Musk oxen—cow, calf, and bull.* (J. J. Teal, Jr.,
Musk Ox Project, Alaska)

experts had proved that mechanically processed quiviut yarns lend themselves best to knitting, a basic skill already familiar to the Eskimo women.
This development is being furthered under the Musk Ox Project with
the establishment of the Musk Ox Producers' Co-operative, which is
expected to assume ownership of the first village herds and to market
the knitters' work. A few items are already being produced and sold
on an individual order basis to appreciative buyers in the contiguous 48
states. The Nieman-Marcus store in Dallas, Texas, featured Eskimo-
knitted 7-ft. quiviut scarves among the "Incomparables" listed in its
1970 Christmas catalog.

CHARACTERISTICS. The adult male musk ox weighs about 900 lb.,
the females less. Yoke-shaped horns are present in both sexes. The animals are low-set; distinctly higher at the withers; and have a long,
shaggy, dark brown outer hair coat. The very fine, dense, long-fibered
undercoat is a lighter brown color. The tail is short and covered with
shaggy hair like the rest of the body.

The cow has a gestation period of about 8 months and usually produces a single calf. Multiple births are rare. The mammary gland has
four teats and produces a small amount of rich milk.

Calves captured in the wild soon become quite tame, well adjusted
to fenced areas, and are intelligent and easy to handle. They grow rapidly under good feeding and management conditions. A group reared in
Vermont averaged 300 lb. at 10½ mo. of age, roughly one-third larger

than their Alaskan counterparts. Wild herds normally subsist on a variety of arctic vegetation. Musk oxen exhibit a browsing propensity similar to goats and members of the deer family, being especially fond of willow shoots.

Earlier and more frequent calving can be expected under domestication than exists in the wild, where the cow normally produces a calf every other year. Seasonal breeding occurs during the autumn months.

The product yields of musk oxen have not been evaluated to any extent under controlled breeding and management conditions. The meat is similar to beef but is of secondary importance to quiviut production. Yields of quiviut from those kept in Vermont averaged about 6 lb. per adult animal. The character of this fiber is indicated by the statement that 1 lb. of quiviut spun in a 40-strand thread will produce 25 miles of thread.

No distinct breeds of musk ox exist as such. However, over their entire northern breeding range in the wild state a white-faced type native to Greenland differs in that respect from the larger solid-brown type of the same species in the western arctic range. Important objectives with the domesticated herds will be selective breeding, nutrition, and management to increase the quantity and quality of quiviut production and other traits contributing to their usefulness in the economy of the Far North.

READING LIST

Dutilly, Artheme. 1949. *A Bibliography of Reindeer, Caribou, and Musk Ox.* Environmental Protection Section, Office of Quartermaster General Dept. of the Army, Research and Development Branch, Report No. 129.

Mathiessen, Peter. 1967. *Oomingmak.* Hastings House, New York.

Moody, Paul A. 1958. Serological evidence of the relationships of the musk ox. *Journal of Mammalogy* 39(4):554–59.

Teal, J. J., Jr. 1958. Golden fleece of the arctic. *Atlantic Monthly* 201(3):76.

Tener, J. S. 1965. *Musk Oxen in Canada.* Queen's Controller and Printer of Stationery, Ottawa, Canada.

Tietz, W. J., and J. J. Teal, Jr. 1967. Chromosome number of the musk ox. *Canadian Journal of Zoology* 45:235.

[CHAPTER NINE] Llamas and Alpacas [TYLOPODS]

THE SUBORDER TYLOPODA of the order Artiodactyla consists of two genera, *Camelus* (camel and dromedary) and *Lama* (llama, alpaca, vicuña, and guanaco). In essence the tylopods are ruminating animals very much like cattle, sheep, and goats, though the structure of the rumen differs somewhat in having a number of saclike subdivisions. Those of the genus *Lama* have no humps and are much smaller than their camel cousins.

The llama and alpaca are the only domesticated large mammals native to the Western Hemisphere. They are found, along with their wild cousins the vicuña and guanaco, in the high Andes Mountains of Bolivia, Peru, Ecuador, Chile, and Argentina at altitudes of 12,000–16,000 ft. above sea level. Their antiquity is attested to by the beautiful fabrics woven on crude hand looms by aborigines predating the Inca culture. The Incas in turn found them indispensable for food, clothing, transport, and religious sacrifice. The guanaco (*Lama huanaca* or *guanicoe*) is thought to be the ancestor of the llama and alpaca. The vicuña (*Lama vicugna*) is the smallest of the group and produces one of the world's finest fibers, being less than half the diameter of the finest sheep's wool. Fabrics woven from vicuña wool sell for upwards of $80 a yard in South America. The highly prized wool has resulted in decimation of the once-abundant herds by hunters. Recently the governments have taken steps to protect them. A notable effort to propagate vicuñas and cross them with the domesticated alpaca is the Cala Cala ranch in Peru (a 3-century-old establishment enclosed by 31 miles of high stone fence) where the animals are kept under captivity in a semidomesticated state.

LLAMAS

The llama (*Lama glama*) is an animal essential to the economy of the Andean natives. Llamas are 4–5 ft. tall at the shoulders and have long, coarse coats ranging from white to black in color. The darker colored ones frequently have white markings around the head, neck,

FIG. 9.1. Llamas in pack train in Peru. Note variation
in color patterns. Bright-colored tassels attached to ears
are not unlike decorative trappings used by horsemen to
dress up their animals. (Photo by Loren A. McIntyre)

ears, and legs. Beneath the outer hairs is the valuable, finer-fibered undercoat. The llama has a short bushy tail, carried gaily; the neck and
ears are carried almost vertically.

An accurate count of llamas throughout the extensive and remote
areas of the Andes is difficult to obtain, but it has been estimated that
there are about 2 million in Bolivia; 1 million in Peru; and another
100,000 in Ecuador, Chile, and Argentina. Perhaps their most important
contribution consists of work as pack animals and production of wool,
although they also provide meat and milk. During the heyday of the
Potosi silver mines in Bolivia as many as 100,000 llamas were used as
pack animals by the Indian miners, who worked as virtual slaves con-
scripted by the Spanish proprietors. Today a Quechua or Aymara Indian
farmer of the altiplano in Peru and Bolivia may keep 10–15 llamas and
perhaps an equal number of alpacas. A llama pack animal can carry a
100-lb. load 15–20 miles a day. The pack trains are stopped to rest and
eat along the way, since llamas seldom graze at night. Llamas and their
kin are said to have the same curious habit as the camel of spitting
saliva at their keeper if overloaded or annoyed in other ways, but they
normally are tractable and reliable work animals. Indian women accompanying pack animals often spin yarn from llama fiber as they walk.

ALPACAS

The alpaca *(Lama pacos)* is somewhat smaller than the llama, standing a little less than 4 ft. high at the shoulders, and is more valued for its fleece. The coat of the alpaca may grow to 30 in. in length—with 10–15 in. being quite common—and it contains no long, coarse, outer fibers like the llama. The alpaca reaches maturity at 2–2½ yr. of age and shears 4–7 lb. of wool. Its color ranges from white to black, similar to the llama. Shearing is done in late November and early December after the seasonal rains. Native fabric production includes ponchos, blankets, and shawls, and much alpaca wool is exported.

It has been estimated that there are over 1 million alpacas in Peru, 250,000 in Bolivia, and about 100,000 elsewhere in their production areas. A few large commercial farms or ranches with Indian herdsmen (pastors) may have 30,000 head of alpacas.

HYBRIDS AND BREEDS

Crosses have been made between the alpaca and llama. The crossbred progeny of a llama male and alpaca female is called a huarizo, and the reciprocal cross is called a misti. Crosses have also been made between the alpaca and vicuña, as on the Cala Cala ranch, in an attempt to combine the size and docility of the alpaca with the fleece quality of the vicuña. The fleece quality of the vicuña has not been obtained in these hybrids. However, the fleece of the alpaca is superior in its own right, the fibers being about three times as strong as fine sheep's wool.

FIG. *9.2. Young alpacas. Fleece hangs nearly to the ground when fully grown.* (Photo by Loren A. McIntyre)

Two breeds of alpaca have been recognized—the Huacaya (or Bacaya) and the Suri. The former is larger and heavier; the latter produces a finer, thicker, heavier fleece. Average fleece weight for the Huacaya is 5½ lb. and for the Suri 6½ lb.

THE FLEECE INDUSTRY

The Lake Titicaca region of Peru and Bolivia is the center of the fleece industry for alpacas, llamas, and their hybrids. Arequipa, in southwestern Peru, is the main collection and shipping center for fleeces. After being sorted for color and type of fiber, the wool is baled in 100 kg. (220 lb.) bales for shipment, usually through the port of Mollendo, Peru. About 85–90% of the world commercial production of fleece from all this group of animals is known as "Arequipa fleece."

Alpaca wool makes up the major portion of the fiber production from this class of animals. In 1935 a total of 8.2 million lb. consisted of 93% alpaca, 4% huarizo and misti hybrid fleeces, and 3% llama. The pigmentation in the fleeces is variable except for vicuña wool, which is uniformly light brown. In wool industry terminology the commercial classification of the wool produced by the alpaca, llama, and their hybrids is:

Class	Spinning Count
Alpaca, coarse	40's
Alpaca, medium	56's
Alpaca, fine	64's

These fascinating and ancient tylopod natives of the Western Hemisphere merit the study of the animal scientist concerned with the welfare of the human society they have so long served. No less intriguing is their interrelationship with a culture stemming from the Inca and pre-Inca aboriginal societies which domesticated them.

READING LIST

McIntyre, Loren. 1966. Flambuoyant is the word for Bolivia. *National Geographic Magazine* 129(2):153.

Rockefeller, Mary and Laurance. 1967. Parks, plans, and people: How South America guards her green legacy. *National Geographic Magazine* 131(1):74.

Stroock, Sylvan I. 1937. *Llamas and Llamaland.* S. Stroock & Co., New York.

Vallenas, P. A. 1965. Some physiological aspects of digestion in the alpaca *(Lama pacos). Physiology of Digestion in the Ruminant.* Butterworth, Washington, D.C.

Weaver, Kenneth F. 1964. The five worlds of Peru. *National Geographic Magazine* 125(2):213.

[CHAPTER TEN] # Dogs [CANINES]

AN'S BEST FRIEND, *Canis familiaris*, has earned its cita-
tion for service by virtue of tenure, devotion, versatility, and universal
distribution. The domestication of the dog dates back at least to the
late Paleolithic or Old Stone Age. The dog was the first animal tamed
by man and probably descended from a doglike animal called Tomarctus
that lived about 15 million years ago. Its forebears are undoubtedly
shared by the wolf, coyote, jackal, and dingo, with which it can interbreed. The true stories of the dog's courage and unwavering devotion
to man are more striking than fiction. No other domesticated animal
exhibits the dog's unique sense of loyalty and obligation in spontaneously coming to man's aid in time of peril. As a genetic subject, the
plasticity of the dog in man's hands is particularly noteworthy, whether
viewed from the standpoint of size, body conformation, coat color and
characteristics, temperament and aptitude traits, or a multitude of other
diverse criteria of appearance or performance. In size alone some representatives of the largest breeds weigh 100 times as much as normal
individuals of the smallest breeds. The dog is distributed among all
countries and peoples of the world. Probably first used by primitive
man to help hunt and capture wild game, the dog at various times
and places has run the gamut of uses man has made of domestic animals.
Furthermore, domestication and husbandry of some other species might
have been substantially delayed without the help of the dog.

In 1928 a German Shepherd became the first seeing-eye dog for the
blind. In 1957 a Russian dog became the world's first space traveler when
sent aloft in an artificial earth satellite by Russian scientists. The study
and description of ova by the Dutch scientist Von Baer, the discovery
of insulin by American scientists Banting and Best, and the classic experiments on conditioned reflexes by the Russian scientist Pavlov are examples of the many scientific advances made possible by dogs as research
subjects. Few in the entertainment world have become better known
than Rin-Tin-Tin and Lassie, whose performances in motion pictures
and television have been viewed by millions. Racing times of Greyhounds compare with the best running horses for comparable distances.
The famed British dog Mick the Miller posted a time of 34 seconds for
600 yards and won over $50,000 in purses during the early 1930s. And
the name of the winning Welsh Terrier Marlu Milady is as familiar to
the Terrier bench-show fraternity as is her namesake in the Jersey cattle
fraternity, to whom the breeder of both these famous matrons passed

along the name of his favorite. Meanwhile, back at countless ranches and farms "old Shep" undergirds the agricultural importance of *Canis familiaris* in America—as do his counterparts throughout the world—in guarding family and possessions, herding livestock, and providing companionship.

THE DOG INDUSTRY

There are an estimated 25 million dogs in the United States. When tractors replaced draft horses in the American economy, dogs and cats consumed virtually the entire draft horse population in the form of commercial pet foods. A very large segment of the veterinary profession is devoted to small animal clinics emphasizing dog health and care.

A dollar value on the extent of the dog industry is difficult to ascertain. Dogs touch every facet of human activity—legal, medical, military, police, agricultural, recreational, industrial, and others.

Mongrels make up the bulk of the dog population. When given freedom of choice, dogs are uninhibited by the biases or prejudices of their human masters and recognize no lines of class, color, appearance, or other barriers. As a sociological observation it can be noted that these animals of mixed ancestry include some of the best adapted, most intelligent, and most useful contributors to the well-being of mankind.

DEVELOPMENT OF TYPES AND BREEDS

The diverse uses of dogs, along with the wide range of environmental conditions under which they are kept, have stimulated the development of many specialized types and breeds. Sturdy, muscular breeds such as the Rottweiler have been used for pulling carts; the powerful, heavy-coated St. Bernard performs rescue work in the Alpine snows; the hardy, fast, and durable Siberian Husky and Alaskan Malamute are adapted to pulling sleds over arctic snow and ice. Hunting dogs to fit the requirements of size, speed, tracking ability, and other traits amenable to the geographic area, terrain, and temperament of their intended quarry have been developed. The water-loving spaniels and retrievers for hunting waterfowl, the pointers and setters for pointing and flushing game birds in land cover, the wily and tenacious coonhound for tracking down and treeing the evasive raccoon, the keen-nosed Bloodhound for meticulous tracking jobs such as in police work, the short-legged Border Terrier and Dachshund for seeking out game in burrows and dens, the swift Greyhound and Whippet for coursing game, and many other specialists have been bred for specific hunting abilities. The quick and lively terriers have the requisites for killing unwanted rodents. Breeds such as the Boxer, German Shepherd, and Doberman

Pinscher have excelled as police, seeing-eye, and guard dogs and were numerous among the 8,500 dogs that served the United States in the "K-9 Corps" during World War II. A well-trained herding dog such as a Collie, Border Collie, Puli, or Old English Sheepdog can replace several men in the handling of livestock. All breeds are suitable as pets and companions, depending on the particular tastes of the owner. In several breeds an apartment-sized miniature variety has been developed to suit the whims of those desiring a smaller-than-standard dog possessing the given breed characteristics. The Miniature Schnauzer, Toy Poodle, and Miniature Pinscher are examples.

The American Kennel Club, which is the chief registry and promotion organization for purebred dogs in the United States, recognizes 116 breeds of purebred dogs. These are divided into the following six groups based on their principal use: sporting—24, hound—19, working—29, terrier—20, toy—15, and nonsporting—9. Additional breeds have established breed standards and are seeking recognition for registry by the AKC.

EVALUATION

A considerable amount of research has been conducted to measure the intelligence, temperament, and aptitude traits of dogs. Most of these studies have been subjective in nature rather than designed experiments to compare breeds and strains or to determine the mode of inheritance of traits. Most comparative performance measurements involve competition of a sporting nature such as herding trials, field trials, track and sled racing, obedience trials, and bench shows.

HERDING TRIALS. The herding of sheep by dogs predates history. The first sheepdog trial was held in Wales in 1873, and organized trials were initiated in Germany about the beginning of the 20th century. Herding trials, often held at regularly scheduled livestock shows, have become popular in Great Britain, Australia, and the United States. Except for some noncompetitive herding exhibitions, the dogs are worked one at a time and are scored on the manner and style of handling the sheep and obedience to their masters in completing the required routine within a specified time without hurrying the sheep. The Border (working) Collie is the principal breed entered in these contests and is also the predominant farm and ranch dog used for working livestock in North America. Interestingly, this breed of black and white dogs, which is blockier and has a broader skull than the show Collie, is not yet an AKC-recognized breed. A well-trained herding dog exhibits a responsiveness to its master's signaled instructions and a deftness in handling livestock that is truly amazing.

FIELD TRIALS. Sporting dog field trials are generally conducted within the following five categories: (1) pointers, (2) Beagles, (3) retrievers, (4) spaniels, and (5) hounds.

The first recorded public field trial was held for bird dogs on May 1, 1866, at Cannock Chase, near Stafford, England. W. Brailsford, the manager and secretary of the trial, has since been referred to as the "Father of field trials." Three judges officiated at this first trial and rated the entries on a 100-point scale according to the following score card:

> Pace and range .... 20 points
> Obedience .... 20 points
> Style in hunting .... 15 points
> Game-finding abilities ...... 20 points
> Style in pointing .... 15 points
> Merit in backing .... 10 points

Opinions on the scoring system were sharply divided, and revisions subsequently led to supplanting the score card system with the heat system. The latter system involved direct competition between bracemates, with the loser of each heat being eliminated from further competition in the trial. An obvious shortcoming of this system was the possibility of eliminating the second best dog in the trial if it happened to get paired in an early heat with the best one. The spotting system, which is still in vogue, was then introduced. This involves the judges' evaluation of each dog in the stake against the balance of the field and will allow dogs competing in the same heat to end up first and second in the trial.

The first grand field trial in the United States was held near Memphis on October 8, 1874, sponsored by the Tennessee State Sportsmen's Association. A black setter named Knight topped the field of nine entries by scoring 88 points on a score card system similar to that used in the first British trial. Some bird dog field trials in the United States are conducted under the rules and regulations of the American Field Publishing Co., Chicago, which maintains the Field Dog Stud Book as an authentic registry for such breeds. The AKC prescribes rules and procedures for field trials held under its auspices for AKC-registered animals. These are set forth for pointing breeds, Dachshunds, retrievers, spaniels, Beagles, and Basset Hounds. The stakes (classes) and recognition awards vary among breeds; the appropriate AKC rules booklets may be consulted for details. In most competitions AKC "points" may be earned toward the permanent designation of Field Trial Champion. Points earned from designated winnings are determined primarily by the number of dogs competing.

Retriever field trials started in 1931, spaniel field trials in 1924, and Beagle trials in 1890. Labradors have outdistanced the nearest competing breed over the years by a margin of three to one in the official nonslip retriever trials. Competitive trials have also been conducted for Foxhounds, Coonhounds, and Bloodhounds but are not provided for in official AKC rules. Formal competition for hounds has been somewhat less significant than for those breeds used in hunting and retrieving game birds. The English sport of "riding to the hounds" has enjoyed some popularity in the past through hunt clubs organized in Maryland, Viriginia, and other areas of the East.

TRACK AND SLED RACING. Dog racing with Greyhounds or Whippets has increased since the development of the mechanical rabbit in 1919. The sport has been especially popular in California and Florida and, like horse racing, usually involves pari-mutuel betting. Many Greyhounds were imported from England and Ireland and crossed with American Greyhounds to meet the demand for increased track racing.

Sled racing with Siberian Huskies, Alaskan Malamutes, or crosses frequently referred to merely as Eskimo dogs is a popular sport in Alaska, Canada, and some states south of the Canadian border. The races vary in distance and number of dogs in the team. Some are run in heats of about 25 miles, and some total distances may run more than 500 miles. Times vary considerably depending on conditions at the time races are scheduled. George Attla, Jr., an Athabascan Indian from Huslia, Alaska, won the 75-mile, three-heat World Championship sled-dog race at Anchorage in 1958 and twice thereafter. In the 1958 race his team set a new heat record of 1 hr., 57 min., 6 sec.—about 26 miles per hour.

DOG SHOWS. Visual appraisal of form and function of purebred dogs is extensively provided for through scheduled shows approved by the AKC and judged by officially licensed judges. Each breed has an official standard against which to evaluate the merit of an individual. Canada and Mexico provide similar competitions. The most prestigious single show for all breeds is the Westminster Kennel Club Show held at Madison Square Garden in New York City in February each year since first inaugurated in 1877—one of the oldest regular sporting events in the United States. Entries of 2,500 dogs are not uncommon at this show.

Dog shows of this type are usually referred to as "bench shows," since the dogs are exhibited in stalls mounted on benches when not being shown in the ring. Professional dog show organizations usually provide all the necessary benching, ring facilities, and management for the sponsoring club. Unbenched shows may be held but are less common. Individual breed groups often hold specialty shows for their particular breed in addition to the all-breed shows held by local kennel clubs. Except for a few major events such as Westminster, most dog shows are scheduled on weekends to cater to the leisure time of spectators and amateur exhibitors. At the regular shows points may be earned by the winning entries toward a permanent Championship title, thus permitting the word "Champion" or its abbreviation to become an official suffix to the dog's registered name. Championship points are recorded for Winners Dog and Winners Bitch according to an AKC scale of points based on the number of dogs of each sex competing in regular classes in the breed or variety. A total of 15 points, accumulated at shows of specified number and point value, with different judges represented, is required for a dog to become a Champion of Record. Recognition as a "Dual Champion" may be gained by a dog that is both a bench show and field trial Champion of Record.

Regular show classes, divided by sex, for each breed or variety are: puppy, novice, bred-by-exhibitor, American-bred, and open. The top placing animals in each of these classes, by sex, are brought together in the Winners Class to determine Winners Dog and Winners Bitch. These, together with dogs that have already accumulated the points for permanent championship, come together to determine Best of Breed. Ultimately, from the Best of Breed winners are selected Best in Group and Best in Show. A miscellaneous class is provided for breeds not yet accorded recognized AKC registry status. Dog owners frequently place show dogs in the hands of professional handlers to be campaigned on the show circuit. The showing of dogs in the ring, especially the working, sporting, and other utilitarian types, is not unlike the halter classes at a horse show.

Obedience classes are becoming increasingly popular at dog shows. Many local kennel clubs hold classes for instruction and practice in obedience training, and 4-H project work with dogs often incorporates this activity along with junior showmanship. Regular show classes for obedience are provided within the novice, open, and utility categories, with specified exercises and score cards for each. The novice exercises and scores will illustrate this competition:

Heel on leash .... 35 points
Stand for examination ...... 30 points
Heel free .... 45 points
Recall .... 30 points
Long sit .... 30 points
Long down .... 30 points
Total   200 points

Official titles awarded for satisfactory performance in the three categories are, respectively: Companion Dog (C.D.), Companion Dog Excellent (C.D.X.), and Utility Dog (U.D.).

Tracking tests may also be offered, with appropriate recognition (T.D.) for performance.

Detailed regulations for registry, dog shows, and obedience trials are set forth in booklets available from American Kennel Club (51 Madison Avenue, New York, N.Y. 10010).

REGISTRATION AND PROMOTION OF PUREBREDS

The American Kennel Club, a nonprofit organization established in 1884, is the official registry organization devoted to the advancement of purebred dogs. Its membership is comprised of 360 independent dog clubs throughout the United States.

The AKC publishes regular volumes of an official studbook, the complete set containing the ancestry records of every dog registered since the club was formed and representing an enrollment of more than 10 million dogs. Dog breeders register litters of puppies, following which the members of the litter are eligible for individual registration.

Another publication of the AKC is the monthly magazine *Pure-Bred Dogs—American Kennel Gazette,* which contains official data on dog activities, informative articles, and advertising.

At its headquarters in New York the AKC maintains a reference library of over 9,500 volumes, including many rare editions, foreign studbooks, and pictures and paintings. The library is open to the public for reference purposes at specified hours during the week.

Dog World, a monthly magazine established in 1916, lays claim to being the world's largest dog magazine and to carrying more paid classified advertising than any other magazine in any field throughout the world. Other published literature on the dog is diverse and voluminous, whether one is interested in either fact or fiction.

The volume of business in AKC registrations alone, together with the numerical rankings of registered dogs, is indicated by the listing of registrations by breeds and groups in 1970 and 1971 (Table 10.1).

TABLE 10.1. Dogs Registered January 1, 1971, to December 31, 1971

	1971		1970	
Breed	Placing	Number registered	Placing	Number registered
Poodles	1	256,491	1	265,879
German Shepherd Dogs	2	111,355	2	109,198
Beagles	3	61,247	4	61,007
Dachshunds	4	60,954	3	61,042
Miniature Schnauzers	5	45,305	5	41,647
St. Bernards	6	35,320	7	27,297
Irish Setters	7	33,516	11	23,357
Labrador Retrievers	8	30,170	10	25,667
Collies	9	28,772	9	26,979
Pekingese	10	27,717	8	27,190
Chihuahuas	11	26,878	6	28,833

TABLE 10.1. *(continued)*

| | 1971 | | 1970 | |
Breed	Placing	Number registered	Placing	Number registered
Cocker Spaniels	12	24,846	12	21,811
Doberman Pinschers	13	23,413	14	18,636
Basset Hounds	14	20,848	13	20,046
Shetland Sheepdogs	15	18,478	15	16,423
Pomeranians	16	17,079	16	16,158
Great Danes	17	16,349	19	13,180
Yorkshire Terriers	18	15,975	17	13,484
Brittany Spaniels	19	15,662	18	13,400
German Shorthaired Pointers	20	14,468	21	12,724
Golden Retrievers	21	13,589	23	11,437
Boston Terriers	22	13,188	20	12,814
Boxers	23	12,617	22	11,483
Scottish Terriers	24	10,765	25	10,248
Old English Sheepdogs	25	10,511	31	6,785
Siberian Huskies	26	10,471	28	7,891
English Springer Spaniels	27	10,076	27	8,945
Fox Terriers	28	9,988	24	10,670
Pugs	29	9,917	26	10,022
Lhasa Apsos	30	9,671	38	6,014
Afghan Hounds	31	8,049	35	6,127
Dalmatians	32	7,883	29	6,961
Cairn Terriers	33	7,738	32	6,698
Samoyeds	34	7,694	34	6,129
Weimaraners	35	7,615	30	6,898
Norwegian Elkhounds	36	7,517	37	6,080
Airedale Terriers	37	6,976	33	6,325
West Highland White Terriers	38	6,754	39	5,801
Bulldogs	39	6,493	36	6,122
Alaskan Malamutes	40	5,621	40	4,373
Maltese	41	4,819	41	4,197
Shih Tzu	42	4,334	42	3,396
Keeshonden	43	3,829	44	3,045
Basenjis	44	3,288	43	3,258
Silky Terriers	45	3,132	45	2,894
Welsh Corgis (Pembroke)	46	2,354	46	2,446
Chow Chows	47	2,352	48	1,813
Vizslas	48	2,242	47	1,973
Chesapeake Bay Retrievers	49	1,928	49	1,611
Newfoundlands	50	1,763	50	1,557
English Setters	51	1,502	54	1,275
Welsh Terriers	52	1,491	51	1,368
Borzois	53	1,445	57	1,138
Schipperkes	54	1,408	52	1,345
Great Pyrenees	55	1,402	58	1,131
Australian Terriers	56	1,313	55	1,210
Miniature Pinschers	57	1,294	53	1,297
Kerry Blue Terriers	58	1,194	56	1,163
Standard Schnauzers	59	1,174	59	1,116
Bloodhounds	60	1,041	62	839
Irish Wolfhounds	61	992	66	769
Gordon Setters	62	937	65	795
Whippets	63	923	61	928
Manchester Terriers	64	888	60	976
Pulik	65	870	64	811

TABLE 10.1. *(continued)*

Breed	1971		1970	
	Placing	Number registered	Placing	Number registered
Italian Greyhounds	66	685	67	699
Bedlington Terriers	67	660	63	813
English Cocker Spaniels	68	653	69	558
Rhodesian Ridgebacks	69	630	68	590
Bull Terriers	70	544	76	398
German Wirehaired Pointers	71	542	73	420
Irish Terriers	71	542	70	529
Bullmastiffs	73	535	72	427
Rottweilers	74	508	71	428
Pointers	75	465	77	396
Belgian Sheepdogs	76	427	79	346
Skye Terriers	77	419	75	405
Japanese Spaniels	78	396	87	285
American Water Spaniels	79	387	81	324
Salukis	80	381	73	420
Papillons	81	369	84	308
Giant Schnauzers	82	367	80	328
Bouviers Des Flandres	82	367	78	348
Norwich Terriers	84	365	89	269
Welsh Corgis (Cardigan)	85	360	85	294
American Staffordshire Terriers	86	340	86	293
Mastiffs	87	325	87	285
Belgian Tervuren	88	319	90	253
Dandie Dinmont Terriers	89	273	83	316
Sealyham Terriers	90	247	82	322
Black & Tan Coonhounds	91	216	92	208
Lakeland Terriers	92	214	93	185
Greyhounds	92	214	94	156
Brussels Griffons	94	205	91	230
Briards	95	153	95	144
Bernese Mountain Dogs	96	152	98	103
Wirehaired Pointing Griffons	97	131	96	112
French Bulldogs	98	110	97	107
Scottish Deerhounds	99	106	101	87
Kuvaszok	100	94	99	102
Affenpinschers	101	82	100	95
Komondorok	102	73	106	54
Foxhounds (American)	103	65	105	57
Flat-Coated Retrievers	104	63	103	69
Irish Water Spaniels	105	60	102	80
Border Terriers	106	59	108	39
English Toy Spaniels	107	43	109	37
Otter Hounds	108	31	110	27
Foxhounds (English)	109	30	115	2
Welsh Springer Spaniels	110	24	112	16
Harriers	111	23	111	19
Curly-Coated Retrievers	112	19	104	59
Clumber Spaniels	113	13	107	42
Field Spaniels	114	12	115	2
Belgian Malinois	115	6	114	3
Sussex Spaniels	116	5	113	4
		1,129,200		1,056,225

TABLE 10.1   *(continued)*

DOGS REGISTERED BY GROUPS

	1971	1970
Sporting breeds	158,925	131,975
Hound breeds	168,000	162,800
Working breeds	291,850	258,550
Terrier breeds	96,075	89,675
Toy breeds	112,925	109,125
Nonsporting breeds	301,425	304,100
	1,129,200	1,056,225

LITTERS REGISTERED BY GROUPS

	1971	1970
Sporting breeds	40,700	36,000
Hound breeds	67,400	68,175
Working breeds	85,275	77,125
Terrier breeds	38,850	38,050
Toy breeds	68,000	68,375
Nonsporting breeds	151,450	158,300
	451,675	446,025

Courtesy of American Kennel Club Inc.

Breed	Principal Uses*	Place of Origin	Height (in.)†	Weight (lb.)‡	Distinguishing Characteristics§
Group 1 Sporting Dogs					
Griffon (Wirehaired Pointing)	Pointing and retrieving game birds	Netherlands & France	19½–23½	50–60	Short, dry, bristly coat; chestnut or combinations of this with steel gray or dirty white. Nose always brown. Straight or gaily carried tail often two-thirds docked.
Pointer	Same as above	Spain, Portugal, Eastern Europe, & England	Not specified	(55–60)	Short, dense, smooth coat with sheen; liver, lemon, black, or orange or combination with white. Full-tapered tail carried straight or level with back.
Pointer (German Shorthaired)	Same as above, including waterfowl; trailing	Germany	21–25	45–70	Short, smooth coat; liver or liver & white; may be ticked. Nose brown and large. Tail always about three-fifths docked.
Pointer (German Wirehaired)	Same as above	Germany	22–26	(55–65)	Same as German Shorthaired Pointer except for wire coat.
Retriever (Chesapeake Bay)	Retrieving, especially ducks	U.S.	21–26	55–75	Eyes wide apart and yellowish color. Coat short, thick, uncurly, with wooly undercoat; dark brown to faded tan or deadgrass.
Retriever (Curly-coated)	Retrieving game, both land and water	England	Not specified	(60–70)	Coat a mass of crisp curls; black or liver. Eyes black or brown, not yellow.
Retriever (Flat-coated)	Same as above	England	Not specified	60–70	Coat dense, flat, fine-textured; black or liver. Eyes dark brown or hazel.

* Nearly all breeds may serve as pets and companions, but this is not specifically indicated if breed has some other major use.
† Unless otherwise indicated, the smaller figure is minimum height for bitches and the larger is maximum height for dogs.
‡ Unless otherwise indicated, the smaller figure represents the lower end of the weight range for bitches and the larger is the upper end of weight range for dogs. Weight is strongly affected by condition, and desired weight is heavily dependent on whether emphasis is for largest size or for miniature size. Where weight is not given in official breed standard, the approximate weight range is shown in parentheses.
§ Not complete breed standards but brief aids to recognition and to call attention to specific points not readily apparent from viewing pictures of the breeds.

TABLE 10.2. (continued)

Breed	Principal Uses*	Place of Origin	Approximate Size		Distinguishing Characteristics§
			Height (in.)†	Weight (lb.)‡	
Retriever (Golden)	Same as above	Scotland	21½–24	60–75	Skull broad, good stop. Eyes dark brown or no lighter than coat. Coat dense and water-repellant with good undercoat; lustrous golden color of various shades; few white hairs permitted on chest.
Retriever (Labrador)	Same as above	Newfoundland	21½–24½	55–75	Skull wide, slight stop. Black or brown eyes preferred, but may be yellow. Coat short, very dense, and without wave; solid black, yellow, or chocolate with small white spot on chest permissible. Nose black or dark.
Setter (English)	Pointing and retrieving game birds. America's oldest gun dog	England	24–25	50–70	Coat flat, good length, no curl; black, white & tan, black & white, blue belton, lemon & white, lemon belton, orange & white, orange belton, liver & white, liver belton, or solid white; allover flecking preferable to color patches. Eyes dark brown.
Setter (Gordon)	Same as above	Scotland	23–27	45–80	Coat soft, shining, straight or slightly waved; black with markings of rich chestnut or mahogany. Eyes dark brown.
Setter (Irish)	Same as above	Ireland	25–27	60–70	Coat mahogany or rich chestnut. Eyes dark to medium brown.
Spaniel (American Water)	Retrieving waterfowl	U.S.	15–18	25–45	Coat closely curled or marcel effect; solid liver or dark chocolate. Eyes hazel, brown, or dark tone, not yellow.

TABLE 10.2. *(continued)*

Breed	Principal Uses*	Place of Origin	Approximate Size		Distinguishing Characteristics§
			Height (in.)†	Weight (lb.)‡	
Spaniel (Brittany)	Pointing and retrieving game birds	France	17½–20½	30–40	Coat dense, flat or wavy, never curly or silky; dark orange & white or liver & white, with some ticking desirable; coat or nose never black. Naturally tailless, or docked to no more than 4 in.
Spaniel (Clumber)	Finding game, especially birds; retrieving	England	Not specified	35–65	Coat silky, straight, very dense; lemon & white or orange & white. Eyes hazel. Nose flesh or cherry.
Spaniel (Cocker) Three color varieties: black, ASCOB, parti-color)	Pet; companion; hunting and retrieving game birds	England	14–15	(22–28)	Coat short and fine on head, flat and silky on body. Eyes black to dark hazel. Nose black to brown, both in keeping with coat color. See official standards for variety coat color specifications.
Spaniel (English Cocker)	Same as above	England	15–17	26–34	Various coat colors. Differs from Cocker Spaniel mainly in larger size.
Spaniel (English Springer)	Hunting and retrieving game birds	England	19–20	(35–50)	Coat liver or black with white markings; liver or black & white with tan markings; blue or liver roan; or white with black, tan, or liver markings. Nose black or liver. Eyes black to dark hazel in keeping with coat color.
Spaniel (Field)	Same as above	England	About 18	35–50	Coat black, liver, golden liver, mahogany red, or roan; may have tan over eyes, on cheeks, feet, and pasterns. Eyes dark hazel to nearly black.
Spaniel (Irish Water)	Retrieving waterfowl	Ireland	21–24	45–65	Long, loose curls forming top-knot between eyes. Coat solid liver. "Rat tail."

TABLE 10.2. *(continued)*

Breed	Principal Uses*	Place of Origin	Approximate Size		Distinguishing Characteristics§
			Height (in.)†	Weight (lb.)‡	
Spaniel (Sussex)	Finding game, especially birds in heavy cover	England	Not specified	35–45	Coat rich golden liver. Nose liver. Tail docked to 5–7 in.
Spaniel (Welsh Springer)	Finding game, especially birds in rough country with heavy cover; retrieving	Wales	Not specified	(33–40)	Coat dark rich red and white; flat and silky, never curly. Nose dark or flesh color.
Vizsla	Hunting and pointing game	Hungary	21–24	(about 50)	Coat short, smooth, dense; no undercoat; solid rusty gold or dark sandy yellow. Nose brown. Tail one-third docked.
Weimaraner	Hunting and pointing game; retrieving, both land and water	Germany	23–27	(55–85)	Coat short, smooth, sleek; solid mouse gray to silver gray. Nose gray. Eyes light amber, gray, or blue-gray. Tail docked to about 6 in.
Group 2 Hounds					
Afghan Hound	Hunting and coursing swift and game	Sinai Peninsula, Egypt 4000–3000 B.C.	24–28	50–60	Head has topknot of long, silky hair. Eyes dark. Nose black. All coat colors permissible, white undesirable.
Basenji	Hunting; companion	Africa about 3400 B.C.	16–17	22–24	Does not bark, but is not mute. Eyes dark hazel. Black nose preferred. Coat short and silky; deep chestnut red, black, or black & tan—all with white feet, chest, and tail tip.
Basset Hound	Hunting foxes, rabbits, pheasants; trailing in dense cover	France	Not over 14	(25–45)	Large skull. Short legs. Muscular. Eyes soft, sad, brown. Short, smooth hair, loose skin; usually black, tan, white, or combination of these.

TABLE 10.2. (continued)

Breed	Principal Uses*	Place of Origin	Approximate Size Height (in.)‡	Weight (lb.)‡	Distinguishing Characteristics§
Beagle (Two varieties: 13 in. height and under and 13–15 in.)	Hunting individually or in packs for rabbits or hare	England, Wales	Note varieties	(18–30)	Eyes large, wide apart, brown or hazel in color. Coat close, hard; any true hound color; usually black, tan, or a combination of these with white.
Bloodhound	Trailing criminals and hunting for lost persons or articles	Middle East 100 B.C.	23–27	80–110	Skin in deep folds around head and neck. Eyes deeply sunken, deep hazel to yellow color. Coat black & tan, red & tan, or tawny.
Borzoi	Coursing game; companion	Russia	26–28 min.	55–105	Small, fine ears. Nose large, black. Eyes dark, soft. Coat long, silky; any color.
Coonhound (Black & Tan)	Tracking and treeing raccoons; hunting other game	England	23–27	(50–60)	No dewclaws. Eyes hazel to dark brown. Coat short, dense; coal black with rich tan above eyes, on sides of muzzle, legs, and breeching, with black pencil markings on toes; no white over 1½ in. in diameter.
Dachshund (Three varieties: long-haired, smooth, wire-haired)	Pet; companion; originally pursuing rabbits, badgers, and other game into burrows	Germany	Not specified	(10–20) Classes provided also for miniature sized animals, under 9 lb. at 12 mo. age or over.	Short legs. Long body. Eyes dark reddish-brown to brownish-black. Coat solid red (tan) of various shades or black with tan points. Note three hair type varieties.
Deerhound (Scottish)	Hunting wolves, coyotes, rabbits; originally hunting deer in Scotland	Scotland	28–32	75–110	Type resembles large, rough-coated Greyhound. Eyes dark. Harsh, wiry coat; dark blue-gray preferred; grays, brindle, yellow, sandy red, or red fawn acceptable.
Foxhound (American)	Fox hunting, singly or in packs	U.S.	21–25	(60–70)	Eyes large, soft, wide apart, brown or hazel in color. Any coat color acceptable, usually a combination of black and tan with white points.

TABLE 10.2. *(continued)*

Breed	Principal Uses*	Place of Origin	Approximate Size		Distinguishing Characteristics§
			Height (in.)†	Weight (lb.)‡	
Foxhound (English)	Fox hunting, generally in packs	England	About 24, with 31" chest girth	(60–75)	Same as American Foxhound.
Greyhound	Coursing hare and jack rabbits; track racing	Egypt 4000–3500 B.C.	Not specified	60–70	Ears small. Eyes dark, bright. General appearance long, tall, streamlined. Coat short, smooth; any color.
Harrier	Hunting hare	France	19–21	(40–50)	Similar to English Foxhound.
Irish Wolfhound	Chasing and killing wolves and coyotes; guard	Ireland	30–32	105–120	Resembles large, powerful, rough-coated Greyhound. Coat gray, brindle, red, black, white, fawn, or any Deerhound color.
Norwegian Elkhound	Hunting big game	Norway 5000–4000 B.C.	19¼–20½ ideals for sexes	(about 50)	Eyes brown. Coat must be gray; shade may vary, and long hairs have black tips. Typical, thick-coated northern dog.
Otter Hound	Developed for hunting otter in England; companion	England	Not specified	(about 65)	Coat color grizzle or sandy, with black and tan rather clearly defined. Resembles Bloodhound, except for long hair.
Rhodesian Ridgeback	Hunting and holding lions and other game at bay	Africa	24–27	65–75 ideals for sexes	Uniquely marked ridge on back. Nose black or brown. Short, sleek coat; light to red wheaten.
Saluki (Also a smooth variety without feathering on ears, tail, and legs.)	Pursuing and killing swift game on sandy or rocky terrain	Egypt 7000–6000 B.C.	23–28 males, females typically considerably smaller.	(about 60)	Eyes dark to hazel color, bright. Nose black or liver. Coat white, cream, fawn, golden, red, grizzle & tan, tricolor, and black & tan.
Whippet	Coursing rabbits; track racing	England	18–22	(18–23)	Eyes large; dark hazel in color. Nose black. Any coat color. Resembles small Greyhound.
Group 3 *Working Dogs* Alaskan Malamute	Sled dogs for freighting or racing	Alaska 1000 B.C.	23–25	75–85	Sturdy, dense-coated, muscular, northern draft-type dog. Coat light gray to black, with white mask and undermarkings; only solid color acceptable is white,

TABLE 10.2 (continued)

Breed	Principal Uses*	Place of Origin	Approximate Size		Distinguishing Characteristics§
			Height (in.)†	Weight (lb.)‡	
Belgian Malinois	Herding livestock; general farm dog	Belgium	22–26	(50–55)	Eyes brown. Nose black. Coat short, straight, with dense undercoat; rich fawn to mahogany with black overlay, mask, and ears; lighter fawn underparts.
Belgian Sheepdog	Herding sheep; police and army service	Belgium	22–26	(55–60)	Eyes brown. Nose and lips black. Coat dense underneath with long guard hairs; solid black.
Belgian Tervuren	Herding livestock; general farm dog	Belgium	22–26	(about 55)	Eyes brown. Coat dense with long guard hairs; rich fawn to russet mahogany with black mask and overlay, black tipped guard hairs, lighter underparts.
Bernese Mountain Dog	Herding livestock; general farm dog	Switzerland 100 B.C.	21–27½	(50–75)	Eyes dark hazel-brown. Coat long, soft, silky, with sheen; jet-black with russet brown or tan on all four legs, over eyes, and each side of white chest; white blaze and points preferred.
Bouviers des Flandres	Farm and watch dog; police and army service	Flanders	22¾–27½	(about 70)	Eyes dark nut-brown; black acceptable. Nose black. Tail docked to about 4 in. Coarse, wiry topcoat; soft, dense undercoat; fawn to black, pepper & salt, gray, and brindle.
Boxer	Guard; companion; police, army, and seeing-eye work	Germany	21–25	(62–75)	Distinctive muzzle and stop, with black mask, except for white markings. Nose black. Eyes dark brown. Coat short, smooth, shiny; fawn or brindle, deep shades preferred; often some white markings, but not on body. Ears cropped. Tail docked.
Briard	Sheep herding; guard, army, and police work	France	22–27	(70–80)	Eyes dark. Ears erect when cropped. 2 dewclaws on each hind leg. Coat long, stiff, strong; any solid color except white.

TABLE 10.2. *(continued)*

Breed	Principal Uses*	Place of Origin	Approximate Size		Distinguishing Characteristics§
			Height (in.)†	Weight (lb.)‡	
Bullmastiff	Guard or watchdog	England	24–27	100–130	Skull large. Nose black. Eyes dark. Coat short, dense; red, fawn, or brindle, with no white markings.
Collie (Rough and smooth varieties)	Herding or driving livestock, general farm and companion dog	Scotland	22–26	50–75	Coat sable & white, tri-color, blue merle, or white. Ear tips should break forward when alert. Note 2 coat varieties.
Doberman Pinscher	Guard; watchdog; police and army service	Germany	24–28	(60–75)	Eyes dark brown to black. Coat black, red, blue, or fawn, with rust markings above each eye and on extremities. Eye color in keeping with coat. Ears cropped. Tail docked.
German Shepherd Dog	Herding livestock; guard; police, army, and seeing-eye work	Germany	22–26	(60–85)	Double coat of medium length. Eyes dark. Nose black. Coat color variable, strong rich colors preferred, no blue or liver color.
Giant Schnauzer	Police service; guard; formerly cattle driving	Bavaria	21½–25½	(about 75)	Eyes dark. Wiry coat; pepper & salt or similar equal mixtures, pure black, or black with tan. Ears usually cropped. Tail docked to three joints.
Great Dane	Guard; companion; originally for hunting large game	Germany	28–30 minimum	120–150	Eyes dark. Ears cropped. Five color varieties: blue, black, fawn, brindle, and harlequin (black splashes over white).
Great Pyrenees	Guard; companion; were used for guarding flocks and for pack and draft service in Pyrenees Mountains	France	25–32	90–125	Eyes dark rich brown. Coat all white or with some markings of badger, gray, or tan; white, dense undercoat.
Komondor	Guardian of livestock; companion	Hungary	23½–25½ minimum	(about 90)	Eyes coffee-brown or darker. Nose black. Coat long, soft, wooly; pure white.

Breed	Principal Uses*	Place of Origin	Approximate Size		Distinguishing Characteristics§
			Height (in.)†	Weight (lb.)‡	
Kuvasz	Guard; companion	Tibet	26 for males, females less.	(165–185)	Eyes dark. Nose black. Coat pure white.
Mastiff	Guard; companion; formerly for fighting	England 55 B.C. or earlier	27½–30 minimum	(165–185)	Eyes brown. Nose black or very dark. Black muzzle and ears. Coat apricot, silver fawn, or dark fawn-brindle.
Newfoundland	Water dog with instinct for lifesaving; guard; companion	Newfoundland	26–28	120–150	Eyes dark brown. Double water-resistant coat; dull jet black, bronze, or black & white. Feet webbed.
Old English Sheepdog	Herding livestock; watchdog	England	22 or more for males, less for females.	(50–65)	Eye color variable in keeping with coat. Nose black. Profuse coat; gray, grizzle blue, or blue-merled, with or without white markings.
Puli	Herding sheep; guardian	Hungary	16–19	(30–35)	Tail usually short, never docked. Nose black. Coat dense, weather-resistant; solid black, rusty-black, or shades of white to gray.
Rottweiler	Driving livestock; pulling carts; guard; police service	Germany	21¾–27	(80–90)	Eyes dark brown. Nose black. Coat short, coarse, flat; black, with tan to mahogany brown markings on cheeks, muzzle, chest, legs, and over each eye.
Samoyed	Watchdog; companion; sledge dog and herding reindeer in Siberia	Northern Siberia 1000 B.C. or earlier	19–23½	(35–60)	Double-coated northern type dog. Coat pure white, white & biscuit, or all biscuit.
Schnauzer (Standard)	Watchdog; companion	Germany	17½–19½	(35–40)	Similar to Giant Schnauzer except for size.
Shetland Sheepdog	Herding sheep; watchdog; companion	Shetland Islands	13–16	(about 16)	Essentially a "pony-sized" Collie.
Siberian Husky	Sled dog	Siberia 1000 B.C. or earlier	20–23½	35–60	Soft, double coat typical of northern dogs; any color or white; usually wolf to silver gray, tan, or black, with white points.

TABLE 10.2. *(continued)*

| Breed | Principal Uses* | Place of Origin | Approximate Size | | Distinguishing Characteristics§ |
			Height (in.)†	Weight (lb.)‡	
St. Bernard (Two coat varieties: rough and smooth)	Guard; companion; Alpine rescue work	Switzerland	25½–27½ minimum	(165–180)	Eyes dark brown. Nose and lips black. Coat color red & white, brindle & white; black mask and specified white markings on extremities.
Welsh Corgi (Cardigan)	Watchdog; companion; driving cattle	Wales 1200 B.C. or earlier	About 12	(15–25)	Nose black. Short, sturdy legs; long body. Coat red, sable, red-brindle, black-brindle, black, tricolor, or blue merle, usually with white markings on extremities.
Welsh Corgi (Pembroke)	Companion	Wales	10–12	18–24	Eyes hazel. Nose black. Short legs; long body. Coat red, sable, fawn, black, or tan, with some white markings acceptable.
Group 4 *Terriers* Airedale Terrier	Guard; hunting; herding livestock; police and army service	England	About 23 for males, females less.	(40–50)	Eyes dark. Nose black. Coat wiry; tan head and darker tan ears; tan legs; tan underside and lower parts of body; black or dark grizzle over top and sides.
Australian Terrier	Companion; hunting small game and vermin	Australia	About 10	12–14	Eyes dark. Nose black. Harsh, straight outercoat; blue-black or silver-black, with rich tan markings on head and legs.
Bedlington Terrier	Pet; formerly hunting badger, fox, otter, etc.	England	15½–16½ ideal	17–23	Eye and nose colors in keeping with coat. Coat a mixture of hard and soft hair; blue, sandy, liver, blue & tan, sandy & tan, or liver & tan.
Border Terrier	Bolting foxes or other small game from burrows or dens; pet	Border of England and Scotland	Not specified	11½–15½	Eyes dark hazel. Nose black. Outercoat wiry; red, grizzle & tan, blue & tan, or wheaten.

TABLE 10.2. (continued)

Breed	Principal Uses*	Place of Origin	Approximate Size		Distinguishing Characteristics§
			Height (in.)†	Weight (lb.)‡	
Bull Terrier (Two varieties: white and colored)	Guard; companion; formerly for dog fighting	England	Not specified	(30–60)	Eyes dark. Nose black. Coat short, flat, harsh; pure white, with markings on head permissible.
Cairn Terrier	Killing vermin; formerly bolting small game from burrows or dens	Scotland	9½–10 ideal	13–14 ideal	Eyes hazel or dark hazel. Nose black. Double-coated, outer coat harsh; any color except white; dark ears, muzzle, and tail tip.
Dandie Dinmont Terrier	Watchdog; killing vermin	England and Scotland	8–11	18–24	Eyes rich dark hazel. Nose and inside of mouth black or dark. Coat 2 in. long and mixture of hard and soft hair; pepper or mustard.
Fox Terrier (Two varieties: smooth and wire)	Watchdog; killing vermin: originally bolting foxes	England	15½ max. for males, females less.	15–19	Eyes dark. Nose black. Coat color predominantly white with black, tan, or black & tan markings.
Irish Terrier	Hunting small game and vermin; watchdog	Ireland	About 18	25–27 ideal	Eyes dark brown. Nose black. Coat dense and wiry; bright red, golden red, red wheaten, or wheaten.
Kerry Blue Terrier	Watchdog; companion; hunting; herding	Ireland	17½–19½	33–40 for males, females less.	Eyes dark. Nose black. Coat soft, dense, and wavy; any shade of blue-gray or gray-blue.
Lakeland Terrier	Watchdog; companion; hunting small game	England	13–15	About 17	Eyes dark. Tail docked. Double coat, outercoat wiry; blue, black, liver, black & tan, blue & tan, red, red grizzle, grizzle & tan, or wheaten.
Manchester Terrier (See also Toy variety)	Companion; killing vermin	England	Not specified	Over 12 and not exceeding 22.	Ears erect; may be cropped to a point. Eyes nearly black. Nose black. Coat short, smooth, glossy; jet black and rich mahogany tan, with tan areas specified.

TABLE 10.2. *(continued)*

Breed	Principal Uses*	Place of Origin	Approximate Size Height (in.)†	Approximate Size Weight (lb.)‡	Distinguishing Characteristics§
Norwich Terrier	Same as above	England	10 ideal	11–12 ideal	Eyes dark, bright, Tail docked. Coat hard and wiry; red, black & tan, or grizzle—all shades.
Schnauzer (Miniature)	Same as above	Germany	12–14 13½ ideal	(about 15)	Same as Standard Schnauzer except for size.
Scottish Terrier	Companion; originally for bolting foxes and vermin	Scotland	About 10	18–22	Eyes very dark. Nose black. Outercoat wiry; steel or iron gray, brindled or grizzled, black, sandy, or wheaten.
Sealyham Terrier	Companion; originally pursuing small game into burrows or dens	Wales	About 10½	20–21	Eyes dark. Nose black. Tail docked. Outercoat wiry; all white or with lemon, tan, or badger markings on head and ears.
Skye Terrier	Companion; originally for bolting foxes	Scotland	9½–10 ideal	(about 25)	Eyes brown. Nose black. Outercoat hard, straight, flat; black, blue, grays, silver, platinum, fawn, or cream.
Staffordshire Terrier	Guardian; companion; originally dog fighting; referred to as Pit Bull Terrier, Pit Dog, etc.	England	17–19	(35–50)	Eyes dark. Nose black. Ears cropped or uncropped. Coat short, close, stiff, and glossy; any color solid, parti, or patched, with desired ratios of mixed colors specified.
Welsh Terrier	Companion; hunting small game	Wales	Males about 15, females less.	About 20	Eyes dark hazel. Nose black. Coat wiry; black & tan or black grizzle & tan.
West Highland White Terrier	Companion; hunting vermin	Scotland	10–11 ideal	(13–19)	Eyes dark. Nose black. Outercoat hard, 2 in. long; pure white only.
Group 5 *Toys*					
Affenpinscher	Pet; companion	Europe	10¼ maximum	(7–8)	Eyes black, brilliant. Nose black. Ears cropped. Tail docked. Coat wiry; black preferred, or may be black with tan markings, red, gray, or other mixtures.

TABLE 10.2. *(continued)*

Breed	Principal Uses*	Place of Origin	Approximate Size		Distinguishing Characteristics§
			Height (in.)†	Weight (lb.)‡	
Chihuahua (Two varieties: smooth and long coat)	Pet; hunting vermin	Mexico	Not specified	1–6 2–4 preferred	Eye and nose color to match coat. Coat soft, length according to variety; any color solid, marked, or splashed.
English Toy Spaniel (Two varieties: King Charles & Ruby, Blenheim & Prince Charles)	Pet; companion	Japan or China, ancient times	Not specified	9–12	Eyes large and dark. Nose black. Coat long, silky, soft. Tail docked. King Charles and Ruby are black & tan and rich chestnut red, respectively; Blenheim is red & white; Prince Charles is tricolor black, white, & tan.
Griffon (Brussels) (Two coat types: rough [wiry] and smooth.)	Pet; companion	Belgium	Not specified	8–10 Not over 12	Eyes black, prominent. Nose black. Ears cropped or uncropped. Tail docked to one-third. Coat reddish brown, black, or specified combinations. No blacks in smooth type.
Italian Greyhound	Pet; companion	Italy 100 B.C.	Not specified	Two classes: 8 & under and over 8.	Coat thin and satiny; red, mouse, blue, cream, white, or all shades of fawn.
Japanese Spaniel	Pet; companion	China, ancient times	Not specified	About 7	Eyes large, dark. Coat straight, long, silky; black & white or red & white (red broadly interpreted).
Maltese	Pet; companion	Malta, 800 B.C.	Not specified	Under 7 4–6 preferred	Eyes dark. Nose black. Coat long, flat, silky, no undercoat; pure white.
Manchester Terrier (Toy) (Two varieties; Note Standard in Terrier Group)	Pet; companion; hunting vermin	England	Not specified	Not over 12 Close to 7 preferred.	Same as Standard variety except for size.
Papillon	Pet; companion	Spain	8–11	(5–11)	Eyes dark. Nose black. Coat long, fine, silky; mostly white, with patches of any color except liver, or tricolor.

TABLE 10.2. (continued)

Breed	Principal Uses*	Place of Origin	Approximate Size		Distinguishing Characteristics§
			Height (in.)†	Weight (lb.)‡	
Pekingese	Pet; companion	China	Not specified	14 maximum	Eyes large, dark, round, prominent, lustrous. Nose black, short, and flat. Outercoat long, flat, with mane and feathering; all colors allowable.
Pinscher (Miniature)	Pet; companion	Germany	10-12½ 11-11½ preferred	(6-10)	Eyes dark. Nose black, except in chocolate-colored dogs. Ears cropped or uncropped. Tail docked. Coat smooth, short, lustrous; solid red or stag red, black with tan markings, solid brown or chocolate with rust or yellow markings.
Pomeranian	Pet; companion; watchdog	Pomerania, Poland	Not specified	3-7 Ideal 4-5	Eyes dark, bright. Outercoat long, straight, glistening, with mane and feathering; 12 permissible colors from black through lighter solid shades to parti-color.
Poodle (Toy) (Three varieties: Note Miniature and Standard in Nonsporting Group)	Pet; companion; trick dog	Germany	10 or under	(under 12)	Same as Miniature and Standard varieties except for size.
Pug	Pet; companion	China	Not specified	14-18	Eyes large, dark, prominent. Head large, round, massive; muzzle short, blunt. Coat fine, short, soft, glossy; silver, apricot-fawn, or black; lighter colors have black mask.
Shih Tzu	Pet; companion	Tibet	10-11	Not over 18 Ideal 9-16	Eyes large, dark, round. Outercoat luxurious, long, dense; all colors permissible.
Silky Terrier	Pet; companion	Australia	9-10	8-10	Eyes small, dark. Nose black. Tail docked. Coat long, flat, silky; blue with specified tan markings.

Breed	Principal Uses*	Place of Origin	Approximate Size		Distinguishing Characteristics§
			Height (in.)†	Weight (lb.)‡	
Yorkshire Terrier	Pet; companion	England	Not specified	Not over 7	Eyes dark. Nose black. Tail docked. Coat glossy, fine, silky, straight; blue on back of neck, body, and tail, with specified tan extremities.
Group 6 *Nonsporting Dogs*					
Boston Terrier	Pet; companion	Boston, Mass.	Not specified	Not over 25. Divided into weight classes for showing.	Eyes wide apart, large, round, dark. Muzzle short, square. Nose black. Ears cropped or uncropped. Short tail, straight or screw. Coat short, smooth; brindle with white preferred, or black with white markings.
Bulldog	Pet; companion; guard; formerly dog fighting and bull baiting	England	Not specified	40–50	Typical undershot bulldog jaw. Very large skull. Nose black. Tail short, straight or screw. Color in order preferred: (1) red brindle, (2) all other brindles, (3) solid white, (4) solid red, fawn, or fallow, (5) piebald, (6) inferior qualities of all foregoing.
Chow Chow	Pet; companion; guard; sporting dog in China	China, 150 B.C.	Not specified	(50–60)	Eyes dark. Nose black. Tongue blue-black and insides of mouth essentially black. Double coat similar to northern dogs; any solid color.
Dalmatian	Companion; sport; formerly coach dog; "firehouse dog"	Dalmatia, Austria	19–23	(40–50)	Nose black in black-spotted dogs, brown in liver-spotted dogs. Coat short, hard, glossy; base color white with distinct black spots or distinct liver-brown spots.
French Bulldog	Pet; companion; watchdog	France	Not specified	Lightweight under 22. Heavyweight 22–28.	Eyes dark. Tail short, straight or screw. Coat smooth, short, brilliant; all brindle, fawn, white, or brindle & white; some other colors do not disqualify.

TABLE 10.2. (continued)

Breed	Principal Uses*	Place of Origin	Approximate Size		Distinguishing Characteristics§
			Height (in.)†	Weight (lb.)‡	
Keeshond	Pet; companion; watchdog	Holland	17–18 ideal	(35–40)	Distinctive "spectacle" lines around eyes; eyes dark brown. Outercoat long, harsh, straight; mixture of gray and black (wolf gray).
Lhasa Apso	Guard; pet; companion	Tibet	Males 10–11, females less.	About 15	Eyes dark brown. Nose black. Coat heavy, straight, hard, with feathering; golden, sandy, honey, dark grizzle, slate, smoke, parti-color—in that order of preference.
Poodle (Miniature & Standard; note Toy in Toy Group)	Pet; companion; trick dog; retrieving; watchdog	Germany	Miniature 10–15 Standard over 15	(Miniature 12–20) (Standard over 20)	Eyes dark. Tail docked. Coat profuse, harsh, and dense; color even and solid at base—blue, gray, silver, brown, cafe-au-lait, apricot, or cream. Several styles of coat clips.
Schipperke	Watchdog; hunting vermin; pet; companion	Belgium	Not specified	Up to 18	Eyes dark brown. Nose black. Tail docked. Coat solid black.

Purebreds not registered by AKC but eligible to be listed and shown in *Miscellaneous Class* include: Akita, Australian Cattle Dog, Australian Kelpie, Border Collie, Cavalier King Charles Spaniel, Ibizan Hound, Miniature Bull Terrier, Soft-coated Wheaten Terrier, Spinoni Italian, and Tibetan Terrier.

Pointer

Wire-haired Pointing Griffon

German Shorthaired Pointer

German Wirehaired Pointer

Chesapeake Bay Retriever

Labrador Retriever

Flat-coated Retriever

Golden Retriever

Curly-coated Retriever

Gordon Setter

English Setter

American Water Spaniel

Irish Setter

Brittany Spaniel

Clumber Spaniel

Cocker Spaniel

English Cocker Spaniel

English Springer Spaniel

Field Spaniel

Sussex Spaniel

Irish Water Spaniel

Welsh Springer Spaniel

Vizsla

Weimaraner

Basenji

Afghan Hound

Beagle

Basset Hound

Bloodhound

Borzoi

Black and Tan Coonhound

American Foxhound

English Foxhound

Smooth-haired Dachshund

Long-haired Dachshund

Wire-haired Dachshund

Scottish Deerhound

Greyhound

Harrier

Irish Wolfhound

Norwegian Elkhound

Otter Hound

Rhodesian Ridgeback

Saluki

Whippet

Alaskan Malamute

Briard

Bernese Mountain Dog

Bouvier des Flandres

Boxer

Smooth Collie

Great Pyrenees

Collie

Bullmastiff

Belgian Tervuren

Doberman Pinscher

Belgian Sheepdog

German Shepherd Dog

Great Dane

Komondorok

Giant Schnauzer

Kuvasz

Mastiff

Old English Sheepdog

Newfoundland

Puli

Rottweiler

St. Bernard

Shetland Sheepdog

Samoyed

Standard Schnauzer

Siberian Husky

Pembroke Welsh Corgi

Cardigan Welsh Corgi

Dandie Dinmont Terrier

Airedale Terrier

Bedlington Terrier

Border Terrier

White Bull Terrier

Colored Bull Terrier

Australian Terrier

Cairn Terrier

Irish Terrier

Smooth Fox Terrier

Wire Fox Terrier

Lakeland Terrier

Kerry Blue Terrier

Manchester Terrier

Miniature Schnauzer

Norwich Terrier

Scottish Terrier

Sealyham Terrier

Skye Terrier

Welsh Terrier

West Highland White Terrier

Staffordshire Terrier

Affenpinscher

English Toy Spaniel

Italian Greyhound

Smooth-coated Chihuahua

Long-coated Chihuahua

Brussels Griffon

Maltese

Japanese Spaniel

Papillon

Silky Terrier

Pekingese

Pomeranian

Miniature Pinscher

Dalmatian

Yorkshire Terrier

Manchester Terrier (Toy)

Boston Terrier

Chow Chow

Bulldog

Pug

French Bulldog

Keeshond

Poodle

Schipperke

Lhasa Apso

READING LIST

American Kennel Club. 1968. *The Complete Dog Book.* Doubleday & Co.,
    Garden City, N.Y.
Dawson, W. M. 1937. Heredity in the dog. *USDA Yearbook of Agriculture,
    1937.* U.S. Government Printing Office, Washington, D.C.
Dog World (monthly). Judy-Berner Publishing Co., Westchester, Ill.

[CHAPTER ELEVEN] Cats [FELINES]

THE CONTRIBUTION of *Felis domesticus* to animal agriculture has gone without notice in most treatises on animal husbandry. Yet the ubiquitous cat—in palace or hovel, ports or ships at sea, farmstead or city street—has long existed in haughty symbiosis with man. Human welfare has been enhanced by the cat in many ways: as a predator on harmful rodents, as a tool in scientific education and research, and as a household pet and companion. While cats seek out the companionship of man to a higher degree than most domesticated animals, they do so on their own terms and generally respond only to such training as suits their whims.

According to a 1961 survey by the Pet Food Institute, about 25% of all American families possess one or more cats. This estimated total of 22 million cats has been accelerating rapidly and does not include the many additional cats to which no one lays claim, such as those inhabiting farm buildings, warehouses, and city streets. By far the preponderant portion of this number consists of the random mating "alley cat" familiar to all. Even these unpedigreed cats belong to a recognized breed—the Domestic or American Shorthair—some representatives of which have been carefully bred for show-ring characteristics.

As is true for many of our breeds and varieties of domestic animals, the origin of the Domestic Shorthair is not well documented. It seems probable from fossil studies that members of the cat family as seen today began to appear about 40 million years ago. The first cats to live with man were probably domesticated by the Egyptians, who revered them and worshipped them as gods. As early as 4,000 years ago, Egyptians carved wooden figures of cats and shaped furniture and jewelry in the form of cats. Mummies of cats as well as of rats and mice placed in the graves with them have been unearthed from ancient Egyptian cat burial sites. On the practical side, the Egyptian cats served to check rats and mice from overrunning the large grain storehouses.

Phoenician traders probably brought Egyptian cats to the European continent between 600 and 900 B.C., where they crossed with native wild types. There is ample evidence that the new breed flourished and thereafter lived in close proximity to man, arriving in North America with the early settlers to form the foundation of the common cat population on this continent. The value of the common cat as a predator on vermin is attested to by a 1970 news release from Colombia, South America. An

FIG. *11.1. Angora.* (Courtesy Mrs. George Thornton)

unexplained decline in the cat population of that country and the concurrent increase of rats and mice resulted in many child deaths attributed to rat bites, especially in the tropical lowlands. Common cats were selling for $12 apiece, and officials were considering large-scale importations of cats to check the vermin increases.

PEDIGREED TYPES AND BREEDS

While the utilitarian value of the cat (also some undesirable activity) resides mainly in the large population of random mating types, a flourishing cat fancy devoted to pedigreed pet and show-ring types exists. The Cat Fanciers' Association, Inc. (P.O. Box 430, Red Bank, N.J. 07701) is the largest of six national cat registry organizations in the United States. The association was formed in 1908 and has a membership of about 350 local clubs throughout the United States and Canada. During its first 50 years of operation it had registered a total of more than 300,000 cats. It was reported in 1970 that the association was registering cats at the rate of over 30,000 per year, reflecting annual increases during the 1960s of over 15%.

The registration of pedigreed cats is very similar to that of dogs and other litter-bearing animals, involving first the recording of the litter which in turn permits members of the litter to be individually registered. The CFA classifies breeds for registration purposes into four groups: natural, established, mutations, and hybrids. The natural breeds form the basis for the other groups. An established breed is defined as having been developed from crossing two or more specified natural breeds. The distinction between an established breed and a hybrid is principally the number of elapsed generations since the natural breeds were crossed to produce it. The close ancestors of an individual of an established breed must be specimens of the established breed, whereas a hybrid's parents may be representatives of different natural breeds (i.e. the hybrids are F_1 individuals). A mutation carries the generally accepted connotation of having spontaneously arisen from a natural breed and exhibits certain unique characteristics not typical of the original parental stock. These may be continued either by interbreeding the mutants or from natural breed parents possessing the mutant gene. The CFA lists 11 natural breeds, 3 established breeds, 3 mutations, and 8 hybrids in their 1970 classification.

CAT SHOWS

Although research studies with cats have been far reaching and involve genetic, physiological, intelligence, and other utilitarian aspects, no organized programs have existed for measuring the performance capabilities of cats. Visual appraisal through a well-organized system of licensed shows has been the main criterion of evaluation. Admittedly, this activity involves emphasis on many fancy points, and thus it suffers the same shortcomings that have made the show-ring a relatively ineffective tool for improving the real merit of most domesticated species. Show standards and show-ring competition do, however, in common with other species, motivate members of the fancy to seek genetic change and improvement.

Show classes and awards in cat shows are similar to those for dog shows. Classes are available for each breed and color within the breed. One exception is the provision of special awards to castrated males

FIG. *11.2. Birman, Seal Point variety.* (Courtesy Mrs. Ralph Griswold)

(neuters) and spayed females (spays). These are designated as Premier-ships, compared to Championships for similar winnings of unaltered cats.

Score cards used as guides in visual appraisal vary among breeds, but the following breakdown for the American Shorthair as set forth in the 1970 CFA Show Standards will illustrate some of the characteristics evaluated in judging cats:

Head (including size and shape of eyes, ear shape and set) .... 30 points
Type (including shape, size, bone, and length of tail) .... 20 points
Coat .... 10 points
Condition .... 10 points
Color .... 20 points
Eye Color .... 10 points

Total 100 points

Accompanying each scale of points is a description of the desired characteristics.

A summary of breeds as specified by the CFA "Classification and Eligibility for Registration" listing most of the recognized types found on the North American continent is presented in Table 11.1. Readers interested in detailed specifications and show-ring standards for these, including numerous coat and eye color variants within some breeds, may obtain such information from the CFA or other cat fanciers associations.

FIG. *11.3. Maine Coon, Blue variety.* (Courtesy Mrs. John B. Rich)

FIG. *11.4. Siamese, Seal Point variety.* (Courtesy Mrs. Stanley Hargrove)

READING LIST

International Cat Fancy (bimonthly magazine). New York.
Suehsdorf, Adolph, and Walter Chandoha. 1964. The cats in our lives. *National Geographic Magazine* 125(4):507–41.

TABLE 11.1. Summary of Cat Breeds

Breed	Place of Origin	Distinguishing Characteristics
Natural		
Abyssinian	Ethiopia (Abyssinia)	Shorthair. Ruddy and red color varieties. Warm reddish-brown surface color most popular.
American Shorthair	U.S., from European background	Shorthair. Several varieties based on coat color, eye color, and special features.
Angora	Turkey	Longhair. Pure white.
Birman	Burma (sacred cat of Burma)	Longhair. Colors include seal, blue, chocolate, and lilac, with typical colorpoint pattern.
Maine Coon Cat	Northeastern U.S. (domestic variant)	Longhair. Several colors or all-white. Long neck and body.
Manx	Isle of Man	Shorthair. Tailless. Short body, high rump. Long hind legs cause a hopping, rabbitlike gait. Many color varieties.
Persian	Iran and Afghanistan	Longhair. Many color varieties within five major divisions.
Russian Blue	Europe and England	Shorthair. Distinctive fine, soft-textured fur with gray or slate blue surface. Large, muscular, long-bodied.
Siamese	Thailand (Siam) (ancient "royal cat of Siam")	Shorthair. Kittens born white, developing typical color pattern with distinctive dark points (extremities) as they mature. Seal, blue, chocolate, and lilac varieties within the basic seal-point and blue-point color types.
Established (Parents must be of the established breed, and only this breed or natural breeds used to produce the breed may appear anywhere in the pedigree.)		
Balinese	Balinese or Siamese acceptable in pedigree. Oriental origin, probably Thailand.	Longhair. Colorpoint coat pattern includes seal, chocolate, blue, and lilac varieties.
Burmese	U.S. from stock brought from Burma about 1930. Related to Siamese. Only Burmese acceptable in pedigree.	Shorthair. Solid seal-brown coat with lighter shades on chest and underbody.
Korat	Thailand. Only Korat acceptable in pedigree.	Shorthair. Blue-gray coat similar to Russian Blue.

TABLE 11.1. (continued)

Breed	Place of Origin	Distinguishing Characteristics
Mutations (Mutations arising from natural breeds)		
Rex	England, from a mutation first seen in 1943. Pedigree may include Rex and/or American Shorthair.	Shorthair. Distinctive soft, curly coat. Same colors as American Shorthair.
Wirehair	Mutant of American Shorthair, which is acceptable in pedigree.	Shorthair. Wiry coat, similar to that of wire-haired dog breeds.
Hybrids (Specified types resulting from combining two or more natural breeds. May be F_1 or other combinations.)		
Colorpoint Shorthair		Siamese type color pattern produced by crossing Siamese and American Shorthair.
Exotic Shorthair		The result of crossing Persians with any shorthair breed(s).
Havana Brown		Shorthair. Produced from Burmese, Siamese, or Russian Blue combinations.
Himalayan		Longhair type produced by crossing Siamese and Persian. Six color varieties displaying dark points typical of Siamese.
Lilac Foreign Shorthair		Produced from crossing Siamese and Havana Brown; consequently may also contain Burmese and Russian Blue inheritance.
Manx-Longhair		Tailless Manx type with long hair introduced from the Persian cross with Manx.
Manxamese		Tailless Shorthair type produced by crossing Siamese and Manx.
Ocicat		Shorthair type resulting from crossing Abyssinian and American Shorthair.

[CHAPTER TWELVE] Small Stock and Fur Animals

Several domesticated or semidomesticated animals not encompassed by the usual definition of livestock play important roles in contributing to human needs. These include animals raised for meat, fur, laboratory research, production of medicinals, biological assay work, and pets or hobbies. A few of the more important of these will be discussed.

RABBITS

Domesticated rabbits are being raised by some 250,000 breeders and hobbyists in the United States. Commercial production is concentrated mainly in California. Each year over 40 million lb. rabbit meat are processed for table consumption; 500,000 rabbits are used for medical research; 200,000 are used for brains and lungs to produce thromboplastin, serum for leptospirosis vaccine, etc.; and about 1 million lb. rabbit fur are produced. This fur production accounts for less than 10% of that used in the United States, the rest being imported.

Vast advances have been made in the production of meat type (fryer) rabbits. The current goal is production of 120 lb. meat per doe annually, and 200 lb. is possible. The feed conversion rate of about 4 lb. feed per pound of meat ranks the efficiency of rabbit meat production comparable to chicken broilers, since the dressing percentage for fryer rabbits is 50–55%.

The domestic rabbit *Lepus cuniculus* is a member of the Leporidae family of the order Lagomorpha. The cottontail and the Belgian hare belong to the genera *Silvilagus* and *Oryctolagus* respectively, within the Leporidae family. The domesticated rabbit is a native of Western Europe and across the Mediterranean Sea in Africa, from whence it has spread to most areas of the world.

BREED DEVELOPMENT AND USES. Rabbits are produced for four main purposes: meat, fur, medical research, and exhibition or hobby use. Four distinct fur structures are recognized; normal (in original breeds), Angora (long-haired, wool producer), Rex (a specific short-haired type), and Satin (a new mutation in several colors and types). A considerable amount of genetic research has been done on the inheritance of coat color in rabbits, and the allelomorphic series ranging from the agouti (wild type) color to the albino has been worked out.

The American Rabbit Breeders Association, Inc. (4323 Murray Ave., Pittsburgh, Pa. 15217) is the official registry and promotion organization for domestic rabbits in the United States. Its first convention was held in Kansas City, Mo., in 1918 as "The First Annual Convention of the National Breeders and Fanciers Association."

Registration of purebred rabbits with the ARBA is carried out with the help of designated area registrars who receive completed registry applications and forward them in triplicate to the national office. Individual identification is by means of tattooing in the right ear. A merit rating is established at time of recording, based on ARBA registration of ancestors in the three-generation pedigree and designated by a seal on the registry certificate.

The ARBA sponsors a Youth Department to promote junior project activities with rabbits, sanctions shows, and licenses official judges for a well-organized system of rabbit shows, usually conducted at county or regional agricultural fairs. The show procedures and awards are very similar to those used for dog bench shows. Best of breed, best opposite sex, and best of variety are named; and a permanent Grand Championship Certificate is awarded when three legs are completed toward the award under prescribed conditions. Utility aspects are emphasized in the judging of breeds kept for meat or fur production.

An official guide is published regularly by the ARBA, carrying informative articles on rabbit breeding and production, breed standards, show awards, and other pertinent items. The *1965 Official Guide* describes 30 breeds, some with more than one variety. A brief summary of recognized breeds is given in Table 12.1.

CAVIES (GUINEA PIGS)

The guinea pig, *Cavea porcellus* of the family Caviidae in the order Rodentia, originated in South America and probably was introduced to Europe in the 16th century by Spanish explorers returning from South America. The exhibition or fancy aspects of the several varieties of guinea pigs have traditionally been fostered jointly with the rabbit breeder organizations. There are three recognized breeds, with several varieties: American, Abyssinian (short, harsh, wiry-coated type), and Peruvian (long, silky, dense-coated type).

TABLE 12.1. Summary of Rabbit Breeds

Breed	Origin & Approximate Date of Appearance	Weight (lb.) Female	Male	Distinguishing Characteristics
American	From crosses of existing breeds, 1917.	9–11	8–10	Blue or white coat color. Meat type. Arched back (mandolin).
American Chinchilla	....	10–12	9–10	Distinctive chinchilla color. Fur and meat producer.
American Giant Chinchilla	From foundation doe (Million Dollar Princess) exhibited at Kansas City, Mo., in 1921.	13–16	12–15	Chinchilla color. Meat most important consideration.
Angora	English and French types. Attempts being made to develop an American type by crossing these two.	7 English 8 French	6 English 8 French	Wool (fur) producer. White predominant color.
Belgian Hare	From Belgium, 1888.	6–9	6–9	Reddish chestnut color. Graceful arched top; trim underline. Different genus than others.
Beveren	From blue St. Nicholas Giants in Antwerp, Belgium, in 19th century.	9–11	8–10	White, with brilliant blue eyes. Also a blue variety like the original.
Californian	From crosses, including Himalayan, produced in California starting 1923. Originally called Cochinellas.	8½–10½	8–10	White with black extremities (Himalayan pattern). Meat type.
Champagne d'Argent	France, in mid-19th century.	9½–12	9–11	Black at birth, later becoming silver. Meat type, also popular for exhibition.
Creme d'Argent	France	8½–11	8–10½	"Orange silver" color.
Checkered Giant	Germany, before WW I.	Over 12	Over 11	Black & white or blue & white. Popular fancier's breed, which is main emphasis.
Dutch	Holland, with further development in England.	3½–5½	3½–5½	Several color varieties. White face blaze and white saddle (belt). Short cobby conformation; one of smallest breeds.
English Spot	England, newly developed.	5–8	5–8	6 colors. Strictly sportsman's or fancy type.
Flemish Giant	Belgium. Flemish Specialty Club is oldest in U.S.	Over 13	Over 12	Largest of all breeds. Large dewlap on does. Several colors. Meat type.
Harlequin	France (Formerly known as Japanese rabbit)	7–9	7–9	Distinctive fancy markings of black and golden orange, one ear and side of head in each color (not bilaterally same color), alternating color rings around body. Mainly for exhibition.

TABLE 12.1. *(continued)*

Breed	Origin & Approximate Date of Appearance	Weight (lb.)		Distinguishing Characteristics
		Female	Male	
Havana	Ingen, Holland in 1898. To U.S. in 1916.	Standard type 5–7 both sexes. Heavy-weight type about 9 both sexes.		Brown. A fur and fancier's breed.
Himalayan	Himalayan Mt. area of Asia. Numerous in China and Russia and one of oldest and most widely distributed breeds.	2½–5	2½–5	White tinged with silver-gray at birth, becoming snow white with black extremities (albino type with pigmentation at points of lowest body temperature).
Lilac	From crosses, first appeared in 1922.	6–9	5½–8½	Unique pinkish-dove color. Desirable commercial fryer.
Lop (English & French)	English Lop one of oldest breeds, more popular than French.	10+	9+	Novelty display type with large drooping ears up to 28 in. from tip to tip. Sooty fawn or black & white.
New Zealand	Originated in U.S. in early 1900s.	10–12	9–11	White, black, and red types. Mainly a commercial meat producer. Colored types more important to fanciers.
Palomino	American breed developed from crossing several existing ones. One of newest recognized by ARBA.	9–11	8–10	Lynx and golden color types. Raised mainly as commercial meat producers.
Polish	Probably from Dutch and Himalayan crosses in late 19th century.	3½ and under, both sexes.		Very small, neat, cobby, sprightly breed. 4 color varieties. Strictly a fancy breed, or for magician's use.
Rex	Produced by French farmer Desire Cuillon in 1919 from wild Belgian Hares.	8 +	7 +	Short "rex" type fur and curly whiskers. A mutation used mainly for unique pelt. Several color shades.
Sable (or American Sable)	Mainly from Chinchilla background, accepted as a standard breed in 1931.	8 +	7 +	Principally fur breed, but good size for meat use. Sepia brown with darker extremities.
Satin	American mutant developed from the Havana.	8½–11	8–10½	Recessive mutant with distinctive fur structure and sheen. Several color varieties, white most numerous. Fur and meat breed.
Silver (or English Silver)	Old breed possibly from Siam, long popular in England.	4–7	4–7	Long fostered by fanciers. Short, plump, compact type in 3 varieties. Not numerous in U.S.

TABLE 12.1. *(continued)*

Breed	Origin & Approximate Date of Appearance	Weight (lb.)		Distinguishing Characteristics
		Female	Male	
Silver Fox	From U.S. crossing to simulate silver fox pelt, accepted as standard in 1925.	10–12	9–11	Two colors, black and blue. Evenness of silvering most important. Principally fur type.
Silver Marten	Mutation from Chinchilla rabbit.	7½–9½	6½–8½	Four colors, black most popular. Fur or fancier's type.
Tan	England, about 1887.	4–6	4–5½	Small, compact, fancy breed. Deep reddish-tan undermarkings combined with black, blue, chocolate, or lilac top color.

Since the real significance of the guinea pig to animal agriculture lies in aspects other than the traditional breed emphasis, the present discussion will center around its most important contribution. Millions of these animals are used annually for laboratory experimentation.

Parameters of inestimable value to large animal genetic improvement and experimentation were developed in classic experiments with guinea pigs at the USDA Research Center at Beltsville, Md., by pioneer population geneticist Sewall Wright. Wright's work extended over a quarter-century and involved records on over 100,000 guinea pigs, including 23 separate families, each descended from an original pair exclusively by brother × sister matings; a control group in which inbreeding was avoided; and crosses among the inbred families. This experimentation is detailed in USDA Bulletin 1121, *The Effects of Inbreeding and Cross-breeding on Guinea Pigs.*

The pioneer work of Wright was further elaborated and extended to research on livestock populations by J. L. Lush of Iowa State University, to whom virtually all present-day animal breeding (population genetics) specialists trace their academic specialization. Many of the genetic principles developed by Wright and further refined by Lush have been widely applied to the improvement of farm animals for commercial production.

HAMSTERS

Like guinea pigs, hamsters are very valuable for laboratory work because of their small size, clean habits, and a reproductive rate that is among the most rapid of all mammals. Hamsters belong to the family Muridae of the order Rodentia. The European or Chinese hamster, *Cricetus cricetus,* is larger than the Syrian (golden) hamster, *Cricetus mesocricetus.* In addition to their laboratory use, golden hamsters make interesting pets. Their name comes from the German word *hamstern,* meaning *to hoard,* derived from their habit of cramming food into cheek pouches and later storing it in their dens.

CHINCHILLAS

The chinchilla, named by early Spanish explorers after the Chincha Indians of South America, is a native of the high Andes Mountains of Peru, Bolivia, and Chile. Its lustrous blue-gray fur makes its pelt valuable for coats, or its hair may be used for weaving. The Chinchilla rabbit derived its name from the typical coat color resembling this animal. Three domesticated species of the genus *Chinchilla* are recognized, belonging to the family Chinchillidae of the order Rodentia.

Chinchilla breeding has been heavily promoted and glamorized as a money-making enterprise since the first 11 captured and introduced to California from Chile by a mining engineer, M. F. Chapman, in 1923 successfully established a domesticated foundation. This commercialization of breeding ventures by aggressive promoters has tended to preclude the establishment of a true pelt value in the fur industry. Pelts marketed incidental to the expansion and culling of breeding units have usually sold for about $25, with rather wide individual variations. A full-length coat would require up to 150 pelts, which gives a clue to the luxury price of a finished chinchilla garment. The future place of the chinchilla in the commercial fur industry remains to be seen.

Chinchillas make clean, desirable pets suitable for urban dwellers, in addition to the possible profit from well-managed commercial chinchilla farms. Considerable effort is being expended by serious breeders to improve fur quality through advanced breeding and management methods.

RATS AND MICE

Albino varieties of the rat *(Rattus norvegicus)* and the mouse *(Mus musculus)* are used by the millions annually for laboratory research on cancer, nutrition, genetics, and for biological assay. Carefully maintained inbred stocks or hybrids produced from crossing inbred lines are used for genetic and cancer research, where a known genetic background is essential. These stocks are supplied in the United States by a few specialized commercial producers.

As laboratory animals, domesticated rats and mice have no peer in their direct research contribution to human health.

FERRETS

The domesticated type (or at least those raised in captivity) of ferret is *Mustela furo,* a member of the suborder Fissipeda of the order Carnivora. These small weasel-like animals were once kept by sportsmen for use in flushing wild rabbits from their burrows. Many have also been used in laboratory studies on the physiology of reproduction and lactation, dental health, and disease. Their normal seasonal breeding habits may be altered in captivity by artificial lighting.

MINK

Mink ranching is a sizable business in the northern part of the United States and in Canada. Some 3,000 farms in this country produce over 3 million pelts annually. The mink, *Mustela vision,* is a carnivore belonging to the suborder Fissipeda of the order Carnivora. Large quantities of slaughterhouse by-products as well as horse meat are consumed by ranch mink. It is interesting to note that safeguards surrounding the use of hormone implants for growth stimulation in beef cattle and broiler production were brought under closer scrutiny with the observation that reproductive performance of mink was upset by the feeding of chicken heads containing the implants.

The value of mink pelts was well established at an early date by wild skins marketed by pioneer trappers. The Hudson's Bay Company, long a major dealer in skins, handled about 1,000 annually between 1752 and 1795. Their annual volume first reached the 100,000 mark in 1831, and in 1890 they shipped over 300,000 wild mink skins to London from U.S. and Canadian trappers. The first wild mink trapped for breeding came mainly from Alaska and Quebec, with Yukon mink being added later.

One of the first recorded efforts to domesticate the mink or adapt it to production under captivity was by Patterson Bros., implement manufacturers of Richmond Hill, Ontario, Canada. This occurred about 1870, and the number of mink kept reached well over 100. The Don Head Farms where this took place are still in operation and have achieved fame as a Southdown sheep and Jersey cattle breeding establishment. Don Head Jerseys continue in the limelight at major shows and sales on the North American continent in the present decade.

The first recorded appearance of an important coat color mutant was in 1931. In 1929 and 1931 the first mink shows were held in the United States and Canada respectively. The aforementioned mutant, the platinum or Silverblu, appeared on the ranch of William Whittingham of Arpin, Wis. The mutant individuals reproduced and founded the Whittingham strain. Other strains of the same mutant were subsequently developed by other mink ranchers. Pelt shows started shortly thereafter. Frank Gothier of Anthon, Iowa, was an early judge at such shows and served as chairman of the committee for development of mink of the American Fox and Fur Breeders' Association.

The first offering of Silverblu mink pelts for commercial use was in New York in 1944 and brought the highest price ever paid for ranch mink pelts at a fur auction. The top pelt brought $265 and the 2,500 sold averaged $147.29. Of these, 65 were consigned by Larry Moore of Suamico, Wis., whose success in the development and promotion of mutation mink stimulated his interest in recessive red-and-white Holsteins, which at the time were unwanted castoffs when they appeared in Holstein herds. Starting with mutants appearing in the famous Winterthur Farm herd in Delaware, he developed the first red-and-white herd of national importance.

In addition to the important Silverblu mutation, many other mink colors affected by either dominant or recessive genes have subsequently appeared and been capitalized on. Because of the changing whims of fashion, the search goes on among mink breeders for unique new types which will capture milady's fancy. An indication of the diversity of mutant color types currently recognized is given by the following list of color classes shown at the International Mink Show at Milwaukee, Wis., Jan. 10–12, 1969, where a total of 682 entries were exhibited:

White	Brown, Light-medium
Aleutian	Brown, Light
Blue Iris	Brown, Extra pale
Sapphire, Medium	Brown, Extra extra pale
Sapphire, Light	Brown, Pale
Violet	Heinen
Pearl	Hope
Pearl, Double	Lavender
Pearl, Triple	Standard dark
Opaline	Brown shadow
Pink	Blue shadow
Brown, Medium	

(A BOS [Breath of Spring] variety of several of the above was also shown. The dominant BOS gene can be superimposed on any of the above color types to give a characteristic surface tinge to the fur.)

The continuing significance of the fur industry to animal agriculture is recognized by the opening of a newly built facility for the USDA Fur Animal Experiment Station at Ithaca, N.Y., where research studies dealing with genetics, nutrition, and management of fur animals are conducted. However, demand for natural furs has declined sharply in recent years.

FOXES

The production of fur from silver foxes bred in captivity became established almost concurrently with that of the more highly prized mink. Prior to the development of mink and fox farms the demand for furs had depleted the wild resources of the North American continent. Great fortunes had been amassed in the business during colonial times when wild skins could be purchased for a mere pittance from Indian trappers.

Black (silver) foxes appearing in Canada and Alaska proved to be mutants of the red fox *(Vulpes fulva)*. In 1894 two farmers on Prince Edward Island, Canada, obtained by capture and purchase some cross and black (silver) foxes and started breeding them in captivity. Their secret in being able to produce all desired mutants soon spread to other fox farms on the island. Just how profitable these operations were became known when 1910 sales figures showed 25 pelts averaging $1,339, with a top of $2,627. This initiated a boom in fox farming, which continued until World War I before subsiding briefly and again accelerating in the 1920s. In spite of much lower prices per pelt, the total value of sales of some 200,000 pelts in 1934 was over $7 million. Fox farming, like mink farming, has been successfully conducted throughout northern United States, Canada, the Scandinavian countries, USSR, and some other cool climatic areas of the world. The market for furs is as capricious as that of women's fashions generally. Mink has been the fur of choice in recent years.

DEVELOPMENT OF TRUE BREEDING TYPES. The spontaneous occurrence of silver foxes among the common red foxes of Canada and Alaska gave rise to two genetic types, which were elaborated by subsequent studies. It was found that crossing the Alaskan silver with the standard (Canadian) silver resulted in progeny intermediate in color between the red fox and silver fox, which are referred to as cross foxes. Studies were initiated in 1928 at the U.S. Fur Animal Experiment Station, Saratoga Springs, N.Y., under the direction of Karl B. Hanson of that station and B. L. Warwick of the Texas Agricultural Experiment Station, to investigate the genetic basis of the two black (silver) mutants. They postulated a two-gene theory which was later substantiated. In brief, the genotypes of the different color types are as follows:

AABB — Common red
AABb — Smoky red
AAbb — Standard (Canadian) black
AaBB — Alaskan red cross
AaBb — Blended cross (Alaskan black x Standard black)
Aabb — Substandard black

aaBB — Alaskan black
aaBb — Sub-Alaskan black
aabb — Double black

It should be noted that black and silver are used interchangeably, even though there are different degrees of silvering. In essence the foxes of this type are black with silver-tipped outer guard hairs.

The development of a significant animal enterprise with the silver fox is another example of man's ability to utilize an animal resource and, through husbandry and application of genetic principles, to capitalize on rare natural occurrences such as gene mutations.

Note: All rabbit photos are courtesy of Ivan Miller.

FIG. *12.1. American Chinchilla rabbit.*

FIG. *12.2. Angora.*

FIG. *12.3. Belgian Hare.*

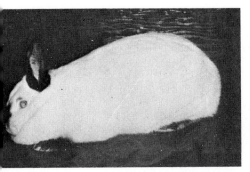

FIG. *12.4. Californian.*

FIG. *12.5. Champagne d'Argent.*

FIG. *12.6. Checkered Giant.*

FIG. *12.7. Dutch.*

FIG. *12.8. English Spot.*

FIG. *12.9. New Zealand Red.*

FIG. *12.10. Palomino* (left) *Golden,* (right) *Lynx.*

FIG. *12.11. Satin (Chocolate).*

FIG. *12.12. Silver Marten.*

Note: All mink photos are courtesy of Larry Moore.

FIG. *12.14. Mink (Silverblu).*
Single recessive gene action
produces this.

FIG. *12.13. Mink (Plain*
Dark).

FIG. *12.15. Mink (Blue*
Frost), due to heterozygous
dominant gene action.

FIG. *12.16.* *Mink (Albino),*
recessive white. Note char-
acteristic sooty smudge on
bridge of nose.

FIG. *12.17.* *Mink (Dominant*
White). Note differences
between this white variety
resulting from a dominant
gene and the albino variety
resulting from a recessive
gene.

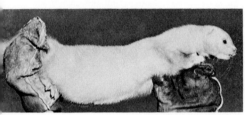

FIG. *12.18.* *Mink (BOS Vio-*
let), a triple recessive mu-
tant. Note tinge of color on
tips of fur.

FIG. *12.19.* *Guinea pigs (Ca-*
vies), American variety.
(Courtesy Miss Terri Dirks)

READING LIST

American Fur Breeder (monthly magazine). Harbrace Publications, Duluth, Minn.

Ashbrook, F. G. 1937. The breeding of fur animals. *USDA Yearbook of Agriculture, 1937.* U.S. Government Printing Office, Washington, D.C.

Hodgson, R. G. 1953. *The Mink Book* (3rd ed.). Fur Trade Journal of Canada, Toronto.

Official Guide of the ARBA (periodic volumes). American Rabbit Breeders Association, Pittsburgh, Pa.

Small Stock Magazine (monthly). Lamoni, Iowa.

Index

Welsh ponies, 119, *126*
Welsh Springer Spaniel dogs, 167, 172, *188*
Welsh Terrier dogs, 159, 166, 180, *197*
Wescott, L. B., 105
Wessex Saddleback swine, 60, *65*
West Highland White Terrier dogs, 166, 180, *197*
Westminster Kennel Club Show, 163
Whippet dogs, 160, 163, 166, 174, *191*
White Holland turkeys, 139, *148*
Wiltshire sides, 50
Wirehair cats, 206
Wirehaired Pointing Griffon dogs, 167, 169, *185*

Wool. *See also* Sheep: alpaca and llama, 158; processing and grading, 69–71
Wright, Sewall, 211
Wyandotte chickens, 135, *141, 142*

Yak, 152
Yorkshire swine, 53, 58, *64*
Yorkshire Terrier dogs, 166, 183, *199*

Zebra, 103
Zebu cattle, 20. *See also* Brahman, Cebú